The Logic Designer's Guidebook

The Logic Designer's Guidebook

E.A. PARR
B.Sc., C.Eng., M.I.E.E.

McGRAW-HILL BOOK COMPANY

New York   St. Louis   San Francisco
Montreal   Toronto

Library of Congress Cataloging in Publication Data

Parr, E. A. (E. Andrew)
    The logic designer's guidebook.

    Includes index.
    1. Logic circuits.   I. Title.
TK7868.L6P37   1984   621.3819′5835   84-7864
ISBN 0-07-048492-9

1234567890   DOCDOC   8987654

ISBN 0-07-048492-9

Printed and bound by R.R. Donnelley and Sons

For
Nicky, Jamie and Simon

The Gates are mine to open
As the Gates are mine to close.

Kipling

Contents

Preface

This book has been written with two major aims. The first is to provide an easy-to-read, but none the less thorough, textbook on digital circuits for use by students and engineers. The second aim is to provide a readily accessible source of data on devices in the TTL and CMOS families. It is hoped that the book will be used as a 'Designer's Handbook' and will spend its days on the designer's bench rather than his bookshelf.

Throughout, the emphasis has been on practicalities (and a chapter on Practical Considerations is included). Too often books on digital design seem not to have heard of MSI devices, and get carried away with minimisation. A good design should minimise ICs & costs (not gates) and be easy to understand and maintain. This emphasis is reflected in the text.

There are two common logic families, CMOS and TTL, which account for some 90 per cent of logic circuits. Details of devices from a less common family (ECL) were regretfully omitted on grounds of length, although the family itself is described on some detail.

Logic symbols are a constant source of controversy. Although a British Standard exists, it appears to have been produced more for the benefit of automatic draughting machines than for human beings. The US MIL-STD used by most IC manufacturers is easier to understand, and has been adopted for this book.

The assistance of Gould Analog Devices and RS Components in providing illustrations is acknowledged with thanks. Special thanks are due to Quarndon Electronics (Semiconductors) Limited of Derby for the use of original negatives prepared by them for their Price List and Catalogue, and to Mullard Limited for providing pinout diagrams for their HE devices, thereby relieving me of the need to produce several hundred drawings. It should be noted that my potted data sheets are my interpretations and not rigorous manufacturers data.

Special thanks are also due to my wife, Alison, not only for doing the typing despite a house move the length of the United Kingdom, but also for tolerating a long backlog of unfinished jobs in the house and garden whilst the book was being written.

Andrew Parr.
Minster,
Isle of Sheppey.

The Logic Designer's Guidebook

CHAPTER 1

Fundamental Theory

1.1 Analog and digital circuits

In fig. 1.1a we have a simple circuit where a voltmeter is connected to the wiper of a potentiometer. By moving the wiper, the voltmeter can be set to any desired voltage between 0 V and the supply voltage Ein.

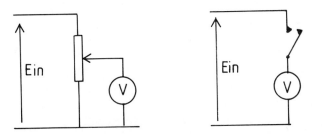

Fig. 1.1 Simple analog and digital circuits. (a) Analog circuit. (b) Digital circuit.

In fig. 1.1b we have another simple circuit, with the voltmeter connected to the supply via a switch. The voltmeter in this circuit can only indicate two values. When the switch is closed, the voltmeter will read Ein; when the switch is open the voltmeter will read 0 V.

There are many ways of classifying the varied aspects of electronics; communications, computing, power electronics, control, and information technology are some of the convenient labels that are attached to different circuits and equipment. Of particular interest to us is the split of electronics into *analog* and *digital* circuits.

Analog circuits are concerned with voltages (or currents) that can take any value (within practical limits). Audio amplifiers, television receivers, DC motor speed controllers are typical analog circuits.

Digital circuits are concerned with voltages (or currents) that can only take certain values. In most of the systems we will consider, the digital circuits will only deal with two voltages. Many circuits operate with voltages that can only be on or off. A light can be on or off, a motor can be running or stopped, a valve can be open or shut.

Figure 1.1a is therefore a simple analog circuit and fig. 1.1b a simple digital circuit.

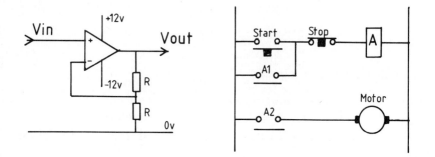

Fig. 1.2 More complex analog and digital circuits. (a) Analog circuit. (b) Digital circuit.

In fig. 1.2a we have a slightly more complicated analog circuit. This is an amplifier of gain 2 (Vout is twice Vin). The output Vout can take any value between +15 volts and −15 volts. If Vin is, say, 4.057 volts, Vout will be 8.114 volts.

In fig. 1.2b we have a common digital circuit. If the start button is pressed, relay A will energise and stay energised via its own contact A1. The contact A2 causes the motor to run. If the stop button is now pressed, relay A will de-energise and the motor will stop.

Typical examples of analog and digital circuits are:

Analog	*Digital*
Potentiometer	Switch
Dial watch	Digital watch
Slide rule	Digital calculator
Voltmeter	Digital multimeter
Variable speed motor	Contactor start/stop motor
Proportional valve	Slam-shut valve
Analog computer	Digital computer

Digital circuits cover a wide range from high current industrial motors to microprocessors. This book is mainly concerned with the design of digital circuits using integrated circuits, although the analytical methods described are equally applicable to most digital systems.

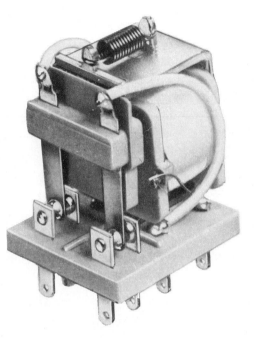

Fig. 1.3 A relay.

1.2 Relay and switching circuits

A relay is a simple and readily understandable digital device. Decision-making circuits based on relays are therefore a useful introduction to the fundamental ideas of logic design.

A typical relay is shown on fig. 1.3, and its internal construction on fig. 1.4a. When a voltage is applied to the coil, the armature is attracted to the coil. The normally open contacts are thus made when the coil is energised, and the normally closed contacts are made when the coil is de-energised. Common symbols for coils and contacts are shown on fig. 1.4b.

A simple relay circuit is shown on fig. 1.5. The relay coil A will be energised when button PB1 is pressed AND switch SW2 is made AND relay B is energised. A series connection of contacts can be said to produce an AND function.

Another simple relay circuit is shown on fig. 1.6. The relay coil A will energise this time if button PB1 is pressed OR button PB2 is pressed OR limit switch LS1 makes. A parallel connection of contacts can be said to produce an OR function.

In fig. 1.7, push button PB1 energises relay A, whose normally closed contact in turn de-energises coil B. The normally closed

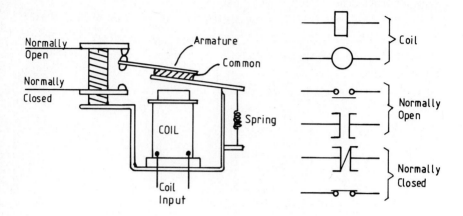

Fig. 1.4 Relay details. (a) Construction. (b) Symbols.

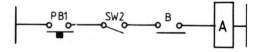

Fig. 1.5 Relay AND circuit.

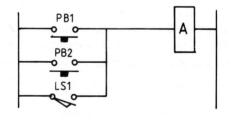

Fig. 1.6 Relay OR circuit.

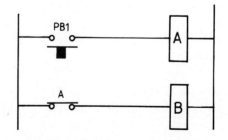

Fig. 1.7 Relay INVERT circuit.

contact can be said to INVERT the state of its coil.

Practical relay circuits are a mixture of AND, OR and INVERT functions. Figure 1.8 shows a control circuit for a motor. The motor is started when relay C is energised. Relay A (whose coil is not shown) selects whether the motor is to be started automatically by relay B or manually by PB1. A local lockoff switch or an overtravel limit switch stops the motor regardless of the state of PB1 or relays A and B.

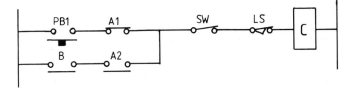

Fig. 1.8 Relay motor circuit.

We could thus say (rather laboriously) that C is energised when:

(((PB1 is pressed) AND (Relay A is de-energised)) OR
((Relay B is energised) AND (Relay A is energised))) AND
(No Lockoff) AND (LS1 NOT OPEN)

This is rather long-winded, even for a simple example. We can shorten it somewhat by using a bar over the top of a coil or contact to indicate the de-energised (or inverted state). $\overline{A}$, for example, is the normally closed contact of relay A. It is verbalised as 'A bar'. We can now write that C is energised when:

$$((\text{PB1 AND } \overline{A}) \text{ OR } (B \text{ AND } A)) \text{ AND } \overline{\text{SW1}} \text{ AND } \overline{\text{LS1}}$$

Another example is shown on fig. 1.9. This is a relay version of the stairwell lighting circuit, where two switches control one light. On fig. 1.9, changing the state of relay A or B will change the state of relay C. If we denote 'C is energised when' by 'C =' we can write:

$$C = (A \text{ AND } \overline{B}) \text{ OR } (\overline{A} \text{ AND } B)$$

This is known as a Boolean equation. Boolean algebra is an analytical method that we shall return to in section 1.4.

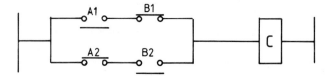

Fig. 1.9 Stairwell lighting circuit.

1.3 Logic gates

We saw in the previous section that the logical functions AND, OR, INVERSION can be obtained with relay circuits. Relays, however, are bulky, consume a lot of power and are unable to switch faster than a few times per second. Electronic circuits performing logical operations are called logic gates.

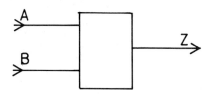

Fig. 1.10 Generalised logic gate.

A simple logic gate can be represented by fig. 1.10. This gate has two inputs denoted A, B, and an output Z. The circuit is designed to work with inputs and outputs that can only have two states. In the TTL logic family, for example, inputs and outputs can be either at a nominal 0 V or a nominal 3.5 V. In the CMOS logic family, signals can be at a nominal 0 V or a nominal 12 V.

These two logic states could be called on-off, high-low, true-false, energised-de-energised (terms that are sometimes found in books and other literature). For simplicity, we shall follow the usual convention and call the higher voltage a '1' and the lower voltage a '0'. This convention is used on most logic families, but the reader should always check definitions carefully when encountering a new logic family.

Let us assume we want to design a logic family with a voltage level of 5 volts being defined as a 1 and 0 volts as a 0. Our first requirement is an AND gate. This will have two inputs A, B and the output Z will be at a 1 when, and only when, both the inputs are 1.

Gate relationships are best shown in tabular form. For the AND gate we have:

Inputs		Output
B	A	Z
0	0	0
0	1	0
1	0	0
1	1	1

This is known as a 'truth table' and defines the gate operations for all possible input states.

Fig. 1.11 Simple AND gate. (a) Circuit. (b) Operation. (c) Symbol.

Consider the circuit of fig. 1.11a. The output Z will always be at the lower voltage of the two inputs. If the inputs can only be at 0 V or 5 V, we can draw up a table as in fig. 1.11b. It can be seen that the output will only be at 5 V when A and B are both at 5 V. The circuit is therefore a simple AND gate. On circuit diagrams, an AND gate is usually represented by the symbol of fig. 1.11c rather than by its electronic components.

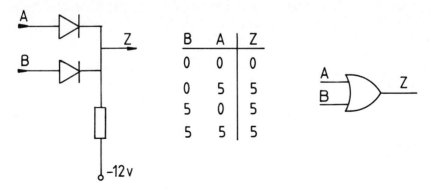

Fig. 1.12 Simple OR gate. (a) Circuit. (b) Operation. (c) Symbol.

In the circuit of fig. 1.12a, Z will be at the higher of the two inputs. As before we can draw up the table in fig. 1.12b showing the output voltages for all possible input voltages. If we adopt our convention of 1 for 5 V and 0 for 0 V we have:

| Inputs | | Output |
B	A	Z
0	0	0
0	1	1
1	0	1
1	1	1

This is the truth table for an OR gate, because Z is a 1 when A is a 1 or B is a 1. The common logic symbol for an OR gate is shown in fig. 1.12c.

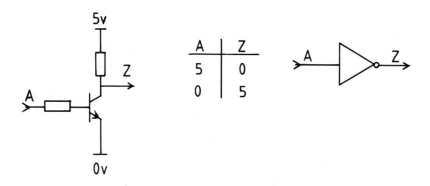

Fig. 1.13 Simple inverter. (a) Circuit. (b) Operation. (c) Symbol.

Our final simple logic gate is shown in fig. 1.13a. When the input A is at 5 V, the transistor will be turned on and the output will be at 0 V. When the input is at 0 V, the transistor will be turned off and the output will be at 5 V. The truth table is therefore:

| Input | Output |
A	Z
0	1
1	0

The output is the inverse of the input, so the circuit is an inverter. The common logic symbol for an inverter is shown in fig. 1.13c.

There are two other gates that can be formed from the circuits of figs 1.11, 1.12 and 1.13.

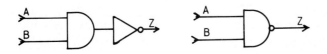

Fig. 1.14 The NAND gate. (a) Function. (b) Symbol.

If we combine the AND gate with the inverter as the circuit of fig. 1.14 we will get the truth table:

Inputs		Output
B	A	Z
0	0	1
0	1	1
1	0	1
1	1	0

This is known as a NAND gate (for NOT-AND).

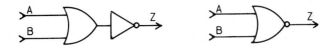

Fig. 1.15 The NOR gate. (a) Function. (b) Symbol.

If we combine the OR gate with the inverter as the circuit of fig. 1.15 we will get the truth table:

Inputs		Output
B	A	Z
0	0	1
0	1	0
1	0	0
1	1	0

This is known as a NOR gate (for NOT-OR).

Note that the logic symbols for the NAND and NOR gates are similar to the symbols for the AND and OR gates with the addition of a small circle at the output. The circle denotes the inversion operation.

In figs 1.11 to 1.15 we have shown AND, OR, NAND, NOR gates with two inputs. These gates can, of course, have any number of inputs. Gates in CMOS and TTL are commonly available in 2, 3, 4 and 8-input forms.

The truth tables for a three-input AND gate and a three-input NOR gate are:

C	B	A	Z AND	C	B	A	Z NOR
0	0	0	0	0	0	0	1
0	0	1	0	0	0	1	0
0	1	0	0	0	1	0	0
0	1	1	0	0	1	1	0
1	0	0	0	1	0	0	0
1	0	1	0	1	0	1	0
1	1	0	0	1	1	0	0
1	1	1	1	1	1	1	0

The truth table for an eight-input gate would have 256 lines!

The circuits shown in figs 1.11 to 1.15 are simple logic gates illustrating AND, OR, INVERSION, NAND and NOR gates, but they have many shortcomings. In particular, the logic levels (0 and 1) are not well defined, and practical considerations such as voltage drops across diodes would soon degrade the levels. In chapter 2 we will describe the common (and some not so common) commercially available families. In chapter 3 we will continue the discussions of logic gates and consider their application in the construction of combinational logic systems.

1.4 Mathematical principles

1.4.1 Introduction

If logical statements are expressed in English, the resultant sentences tend to be cumbersome and difficult to understand. Various techniques have been evolved to display logical expressions in a straightforward, easy-to-follow manner. The most common are truth tables, Boolean algebra and Karnaugh maps.

These techniques also give indications of how a particular logical expression can be converted into a practical circuit using the minimum amount of hardware.

1.4.2 Truth tables

We have already encountered truth tables in section 1.3 when we drew up the truth tables for the AND gate, the OR gate and the inverter. Any simple combination of gates can be represented as a box with one or more outputs. A truth table simply lists every possible input state and shows the corresponding output states.

For example, suppose we have three inputs A, B, C and we require the output to be 1 when two, and only two, inputs are 1. The truth table for the required circuit is:

	Inputs		Output
C	B	A	Z
0	0	0	0
0	0	1	0
0	1	0	0
0	1	1	1 ←
1	0	0	0
1	0	1	1 ←
1	1	0	1 ←
1	1	1	0

With three inputs there are eight possible input states. Three of these have two, and only two, inputs at 1. For these the output Z is 1, for the other five combinations Z is 0.

By looking at the truth table we can see that Z is 1 when

$A = 1$ and $B = 1$ and $C = 0$ (i.e. $\bar{C} = 1$)

OR $A = 1$ and $B = 0$ (i.e. $\bar{B} = 1$) and $C = 0$

OR $A = 0$ (i.e. $\bar{A} = 1$) and $B = 1$ and $C = 1$

The required circuit can therefore be constructed with the logic assembly of fig. 1.16. What the truth table does not tell us directly, however, is whether this is the *minimum* combination of gates to achieve the required output. This is a problem we shall return to later in section 3.3.

The truth table is a convenient way of representing logic statements when there are few inputs. Unfortunately, for large numbers of inputs the tables become unmanageably large. For 3 inputs there are 8 lines on the table, for 4 inputs there are 16 lines, for 5 inputs there are no less than 32 lines and so on. Truth tables are generally used when there are fewer than 5 inputs.

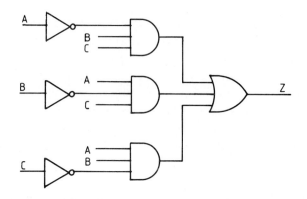

Fig. 1.16 Two, and only two, circuit.

1.4.3 Boolean algebra

In the nineteenth century, a Cambridge mathematician and clergy-man called George Boole devised an algebra to express and manipulate expressions whose components can only have two states. Although originally devised to solve logical problems whose components were 'true' or 'false', it is ideally suited to describe logical relationships in digital circuits. In passing, though, it is worth noting that Boolean algebra is a powerful tool for analysing the 'Mr Brown lives next door to the baker, Mr Smith does not like tomatoes' type of problem.

Boolean algebra can be described and 'proved' in a rigorous mathematical manner. Understanding of the underlying mathematical theory, however, is not necessary (or even desirable) to apply Boolean algebra to digital circuits. All the relevant ideas are self-evident, or can be demonstrated by simple examples.

The AND function in Boolean algebra is represented by the dot (.). We thus have, for example:

$$Z = A . B$$

which means Z is 1 when A is 1 AND B is 1. Sometimes the dot is omitted, allowing the above expression to be written

$$Z = AB$$

The OR function is represented by the addition sign (+). For example, we can write:

$$Z = A + B$$

which means Z is 1 when A is 1 OR B is 1.

The invert function is represented by the bar ‾ as described earlier in section 1.2. The expression:

$$Z = \bar{A}$$

means that Z takes the opposite state to A. When A is 1, Z is 0; when A is 0, Z is 1.

Boolean algebra allows quite complicated logical expressions to be written in a simple form. The circuit shown in fig. 1.8, for example, can be written in Boolean terms as:

$$Z = (\bar{A}.PB1 + A.B) . \overline{SW1} . \overline{LS1}$$

And fig. 1.16 can be written:

$$Z = AB\bar{C} + A\bar{B}C + \bar{A}BC$$

We can now examine the rules governing Boolean algebra. First we have the laws involving one variable, say A:

(a) $A.A = A$
(b) $A + A = A$
(c) $A.1 = A$
(d) $A.0 = 0$
(e) $A + 1 = 1$
(f) $A + 0 = A$
(g) $\bar{\bar{A}} = A$
(h) $A.\bar{A} = 0$
(i) $A + \bar{A} = 1$

These nine laws are rather self-evident, but they are the foundations' of the more useful laws below.

The next two laws are called the commutative laws, and tell us that the order in a Boolean expression is not important:

(j) $A + B = B + A$
(k) $A.B = B.A$

The associative laws allow us to group brackets around variables with the same operators:

(l)　$(A + B) + C = A + (B + C) = A + B + C$

(m)　$(A.B).C = A.(B.C) = A.B.C$

The next two laws are called the absorptive laws, and tell us what happens when the same variable appears with AND and OR operators:

(n)　$A + A.B = A$

(o)　$A.(A + B) = A$

The B variable, it will be noted, has no effect.

The next laws are called the distributive laws, and are useful for factorising Boolean equations:

(p)　$A + B.C = (A + B) . (A + C)$

(q)　$A.(B + C) = A.B + A.C$

Complex Boolean expressions can be expressed in two forms. The first form brackets OR terms, and ANDs the brackets; for example,

$$Z = (A + B) (A + \overline{D}) (B + C)$$

This is known as products of sums, or P of S form.

The second form groups AND terms and ORs the resulting terms; for example,

$$Z = AB\overline{D} + \overline{B}C + A\overline{D}$$

This is known as sum of products, or S of P form.

The complement of a Boolean expression yields the inverse of the expression (e.g. where the expression yields 1, the complement yields 0). Consider the expressions $A + B$, and $\overline{A}.\overline{B}$. These have the truth tables:

	B	A	Z		B	A	Z
$A + B$	0	0	0	$\overline{A}.\overline{B}$	0	0	1
	0	1	1		0	1	0
	1	0	1		1	0	0
	1	1	1		1	1	0

The truth table of $\overline{A}.\overline{B}$ is the inverse of the table for $A + B$, so $A + B$ and $\overline{A}.\overline{B}$ are complementary functions. This is usually written:

(r)　$\overline{A + B} = \overline{A}.\overline{B}$

Similarly, it can be shown that A.B and $\overline{A} + \overline{B}$ are complementary functions. This is usually written:

(s)　$\overline{A.B} = \overline{A} + \overline{B}$

The two expressions (r) and (s) are simplified versions of De

Morgan's theorem which states generally that:

(ri) $\overline{A.B \quad \quad N} = \overline{A} + \overline{B} + \quad + \overline{N}$

(si) $\overline{A + B \quad \quad + N} = \overline{A}.\overline{B}. \quad \quad .\overline{N}$

De Morgan's theorem can be given in a more general form to derive the complement of any Boolean expression in P of S or S of P form. The complement is formed in two stages:

1. Replace each '+' in the original expression by '.' and vice versa;
2. Complement each term in the original expression.

These two stages sound more complex than they really are, and are best illustrated by example. First a simple expression: $\overline{A} + B.C$

Stage 1. Replace '+' by '.' and '.' by '+', giving

$$\overline{A}.(B + C)$$

Stage 2. Complement each term,

$$A.(\overline{B} + \overline{C})$$

so

$$\overline{\overline{A} + B.C} = A.(\overline{B} + \overline{C})$$

We can check this result by constructing a truth table:

C	B	A	$\overline{A} + B.C$	$A.(\overline{B} + \overline{C})$
0	0	0	1	0
0	0	1	0	1
0	1	0	1	0
0	1	1	0	1
1	0	0	1	0
1	0	1	0	1
1	1	0	1	0
1	1	1	1	0

This confirms that $A.(\overline{B} + \overline{C})$ and $\overline{A} + B.C$ are complementary functions.

Now a more complicated expression:

$$(A + \overline{B}).(A + \overline{C} + D).(A + \overline{B}D)$$

Stage 1. Replace '+' by '.' and '.' by '+' giving

$(A.\overline{B}) + (A.\overline{C}.D) + (A.(\overline{B} + D))$

Stage 2. Complement each term

$(\overline{A}.B) + (\overline{A}.C.\overline{D}) + (\overline{A}.(B + \overline{D}))$

The outer brackets are optional in this expression, so

$$\overline{(A + \overline{B}).(A + \overline{C} + D).(A + \overline{BD})} = \overline{A}.B + \overline{A}.C.\overline{D} + \overline{A}.(B+\overline{D})$$

This result could, if required, be checked by a truth table.

Boolean algebra appears, at first sight, to be an interesting academic exercise of little practical use. It does, however, provide a useful tool for reducing the number of logic gates required to implement a given expression. For example, consider the expression:

$$Z = AB\overline{D} + A\overline{B}C + \overline{A}B\overline{D} + AD + \overline{B}\overline{C}\overline{D}$$

Given the four signals ABCD, we need four inverters, four three-input AND gates, one two-input AND gate and a five-input OR gate to implement this directly, a total of ten gates.

The expression can be reduced, using Boolean algebra, to

$$Z = A + \overline{B}\overline{D}$$

This requires two inverters, a two-input AND gate and a two-input OR gate; a saving of six gates over the original arrangement. The reduction also shows that the C variable has no effect.

Minimisation of logical functions is described in detail in section 3.3.2.

1.4.4 Karnaugh maps
A Karnaugh map is an alternative way of presenting a truth table. The map is drawn in two dimensions; two-, three- and four-variable maps are shown in fig. 1.17.

Each square within the map represents one line on the truth table. For example:

square X represents A = 1, B = 0 which can be written $A\overline{B}$
square Y represents A = 0, B = 1, C = 1 which can be written $\overline{A}BC$
square Z represents A = 1, B = 0, C = 1, D = 0 which can be
        written $A\overline{B}C\overline{D}$

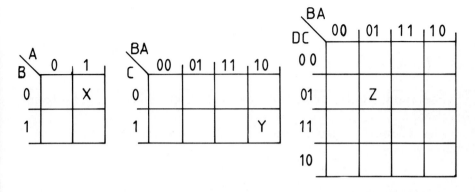

Fig. 1.17 Karnaugh maps. (a) Two-variable map. (b) Three-variable map. (c) Four-variable map.

The essential feature of a Karnaugh map is the way in which the axes are labelled. It will be seen that only one variable changes between horizontally adjacent squares in any row, and only one variable changes between vertically adjacent squares in any column.

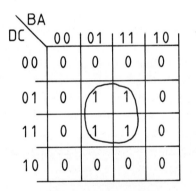

Fig. 1.18 Plot of AC.

The use of this feature is not immediately apparent, but consider fig. 1.18. The truth table contains four terms giving a 1 output. These are:

$$A\bar{B}C\bar{D}, \; ABC\bar{D}, \; A\bar{B}CD, \; ABCD$$

so we could write (quite correctly):

$$Z = A\bar{B}C\bar{D} + ABC\bar{D} + A\bar{B}CD + ABCD$$

18

Examination of the map, however, shows that the D variable and B variable change without affecting the output. The circled squares in fact represent AC, so the above expression can be simplified to

$Z = AC$

This result could, of course, also have been obtained by Boolean algebra.

A Karnaugh map is very powerful technique for minimising expressions, as we shall see in section 3.3.3.

1.5 Specifications

1.5.1 Introduction
Logic gates are not perfect, infinitely fast devices. They have practical limitations, and any successful design must take due consideration of factors such as speed, noise and driving capability.

This section deals with the methods used to specify the performance of logic gates.

1.5.2 Speed
A logic gate, such as the inverter of fig. 1.19a, does not respond instantly to changes at its input. If a perfect pulse is applied to its input, the output will be delayed, and the edges slowed as shown in fig. 1.19b.

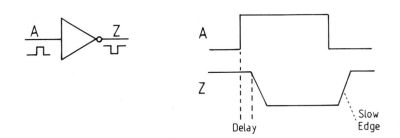

Fig. 1.19 Propagation delay and edge speeds. (a) Circuit. (b) Practical waveforms.

The delay is termed the propagation delay, and is measured from mid-point of the input to mid-point of the output as shown on fig. 1.20. The symbol 'tpd' is used on data sheets. Typical values of tpd are 8 ns for TTL, 80 ns for CMOS, 1 ns for ECL. Propagation delay is the major restriction on the speed at which a digital circuit can operate.

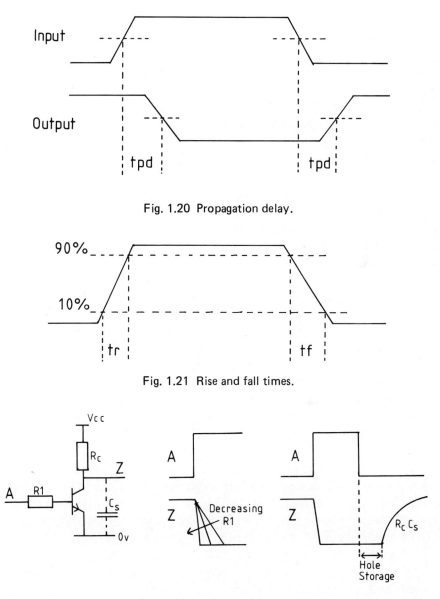

Fig. 1.20 Propagation delay.

Fig. 1.21 Rise and fall times.

Fig. 1.22 Saturating transistor switching. (a) Circuit. (b) Effect of R1 on turn-on. (c) Hole storage and turn-off.

Related to the propagation delay is the time taken for the output to change state. This is known as the rise time (for the 0 to 1 transition) or the fall time (for the 1 to 0 transition). This is usually measured between the 10% and 90% points as shown on fig. 1.21.

The symbols tr and tf (or t_{LH} and t_{HL}) are used on data sheets. Typical values are 5 ns for TTL, 40 ns for CMOS and 1 ns for ECL.

Propagation delay and rise/fall times are interlinked with power consumption. 'Speed' is largely determined by two effects: charge storage and stray capacitance.

With the exception of ECL, most logic common families use saturating transistors to define the logic states close to a supply rail, and to decrease the switching times (i.e. rise and fall times). This means that the base current in fig. 1.22a is much greater than Ic/h_{fe}. This gives the fast-falling edge of fig. 1.22b, as a large collector current is available to discharge the stray capacitance C_s. The more the transistor is overdriven, the shorter the fall time.

When a saturated transistor is turned off, an effect called charge storage (caused by excess minority carriers in the base region of the transistor) delays the turn-off. This turn-off time can be approximated by an exponential return from the collector current equivalent to $h_{fe}.Ib$ as shown on fig. 1.22c. It follows that overdriving a saturated transistor decreases the turn-on time, but increases the charge storage time.

When the transistor does turn off, the stray capacitance is charged via Rc, so the output is an exponential rise of time constant RcCs. This can be reduced by reducing Cs (i.e. keeping the circuit small) or decreasing Rc (which decreases the overdrive, thereby increasing the turn-on time, and at the same time increasing the device power requirements). The designer therefore has to decide on a compromise between propagation delay, edge speeds and power. Normal TTL, for example, has a tpd of 10 ns and a power consumption of 10 mW for a simple gate package. High speed TTL (suffix H) uses lower value resistances and has a propagation delay of 6 ns at the expense of an increased power consumption of 23 mW. Schottky TTL uses a clever trick (described in section 2.6) to avoid charge storage and manages 3 ns propagation delay for a power consumption of 19 mW.

A figure of merit often used in comparing logic families is the speed/power product. This is simply the product of the propagation delay and power for a single gate.

To some extent, however, power consumption and speed/power product are misleading. As speed rises, all the circuit stray capacitances have to be charged and discharged as gate outputs change state. This current is negligible at low speed, but rises significantly as the speed rises above 100 kHz. A CMOS gate, for example, has a power consumption of 0.01 mW. At 100 kHz it consumes 0.1 mW, and at 1 MHz it will consume 1 mW.

High speed and fast edges do not necessarily mean 'good'. As we shall see in sections 11.2 and 13.3, high speed can bring its own problems due to noise and transmission line reflections.

1.5.3 Fan-in/fan-out

The output of a logic gate can only drive a certain maximum load and remain within specification for speed, voltage levels, etc. There is therefore a maximum number of gate inputs that can be driven from a gate output. This is defined by the fan-out and fan-in of a gate.

A simple gate input is called a standard load, and is said to have a fan-in of one. A gate output is defined by how many standard loads it can drive. This is known as the fan-out. A TTL gate output, for example, has a fan-out of ten, so it can drive ten standard gate inputs.

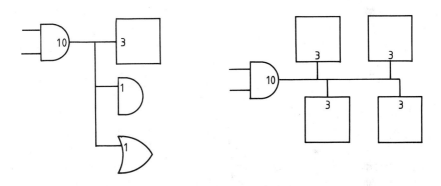

Fig. 1.23 Fan-out/fan-in. (a) Correct. Fan-out 10, fan-in 5. (b) Incorrect. Fan-out 10, fan-in 12.

Some inputs (notably clock inputs) present a load equivalent to more than one standard gate input. These have a fan-in equivalent to a number of equivalent standard loads. The circuit in fig. 1.23a is therefore acceptable (fan-in 5, fan-out 10) but the circuit of fig. 1.23b is unacceptable (fan-in 12, fan-out 10).

1.5.4 Noise immunity

Electrical interference is called 'noise' and can cause 1 signals to appear as 0 signals and vice versa. The ability of a logic gate to reject noise is known as its noise immunity.

To define the noise immunity (or noise margin as it is more commonly termed) we must define the voltage levels in fig. 1.24, where we have a simple AND gate driving another gate. The voltages we shall use are for TTL, but noise margin is defined in a similar way for all logic families.

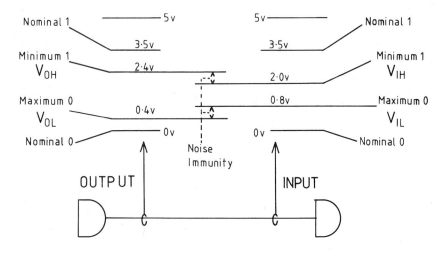

Fig. 1.24 Noise immunity.

First we have the nominal voltages for the 1 state and 0 state; these are 3.5 V and 0 V respectively. Next we define how low the output 1 state can fall (V_{OH}min, 2.4 V) and the maximum to which the 0 state can rise (V_{OL}max, 0.4 V).

Finally we define how low the input 1 state can fall, whilst still ensuring that V_{OH}min of the driven gate does not fall below its specified value of 2.4 V (V_{IH}min, 2.0 V) and the maximum to which the 0 state can rise whilst still ensuring that V_{OL}max of the driven gate does not rise above its specified value of 0.8 V (V_{IL}min, 0.8 V). V_{IH} and V_{IL} are sometimes called the threshold voltages.

With these voltages the DC noise margin is defined as the smaller of ($V_{OH} - V_{IH}$) or ($V_{IL} - V_{OL}$). For TTL, both these expressions yield 0.4 V, so TTL has a DC noise margin of 0.4 V. This result is obtained under the worst conditions of power supply, temperature and load. A more typical figure would be about 1.2 volts.

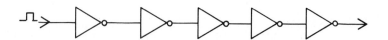

Fig. 1.25 Testing for AC noise immunity.

A figure sometimes specified is the AC noise margin. This is defined as the largest pulse that will not propagate down a chain of gates similar to fig. 1.25. The AC noise margin tends to give a more favourable figure than the DC noise margin, but is probably a more realistic test.

The noise rejection ability of a logic family is not solely determined by the magnitude of the noise margins. Voltage swing, impedances and other factors need to be considered as well. Noise, its causes and its prevention, is discussed in detail in section 13.3.

1.5.5 Transfer characteristics
The transfer characteristic of a logic gate is the relationship between the input and output voltages. Typical transfer characteristics for inverting and non-inverting gates are shown on fig. 1.26.

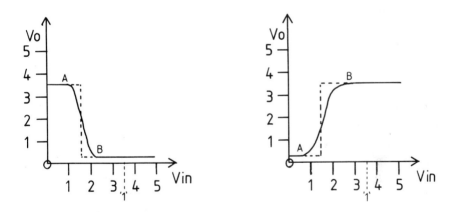

Fig. 1.26 Transfer characteristics. (a) Inverting gate. (b) Non-inverting gate.

The transfer characteristic is closely related to the noise immunity. An ideal transfer characteristic would approximate to the dotted lines. In the region AB the gate is behaving like a linear amplifier. During the time a signal is passing through this region, the device is particularly vulnerable to noise. The more vertical the section AB is, the better the noise immunity. CMOS has a transfer characteristic that is close to the ideal.

1.6 Integrated circuits

With the exception of some industrial logic, all logic families are based on integrated circuits (ICs). This book is not really concerned with the technology of integrated circuit manufacture; suffice it to say that integrated circuits give predictable performance, and their small size gives high speed due to low stray capacitance, etc. The economies of scale applied to IC production give low cost with high volume; in fact common digital ICs are cheaper than individual transistors.

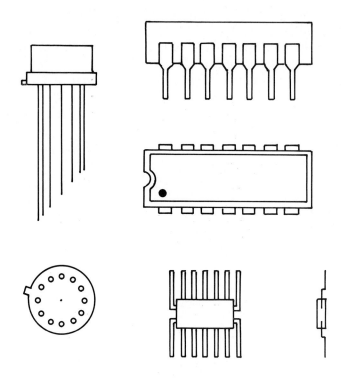

Fig. 1.27 Some IC packages.

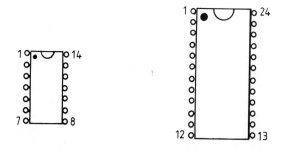

Fig. 1.28 DIL pin configuration (viewed from above). (a) 14-pin. (b) 24-pin.

Three common IC packages are to be found. These, illustrated in fig. 1.27, are the dual-in-line (DIL), the flat pack and the T05. The T05 is largely obsolete, and the flat pack is usually only found in high density applications (it is soldered directly to the PC board on one side only, and does not require holes for the leads. This simplifies track layout). The dual-in-line package is the commonest and cheapest encapsulation. It is available in plastic form (for temperature range 0-70°C) or ceramic form for the full military specification range of −55°C to +125°C.

DIL packages are available in 14, 16, 24 and 40-pin versions. Pin numbering is always anticlockwise viewed from above as shown on fig. 1.28. Pin 1 is to the left of the cut-out, and is sometimes identified by a dot.

CHAPTER 2
Logic Families

2.1 Introduction

In section 1.3 a very simple family of logic gates was described. In this chapter we examine logic families that are commercially available, and some that are historically interesting.

There are logic families for almost every application, from robust industrial systems with 24 volt power supplies and maximum operating speeds of about 100 kHz, to high speed ECL which can operate reliably at over 500 MHz. The main considerations of a designer when choosing a logic family are price, availability, speed, noise immunity and power consumption. Each family described below has different strengths and weaknesses.

2.2 Resistor Transistor Logic (RTL)

RTL is one of the earlier logic families, and is not currently available in IC form. It is, however, the basis of industrial logic families such as NORBIT and NORLOG. RTL exists in two forms, shown in figs 2.1a and b. In each case, if either input goes high the output goes low, giving the truth table (for high = 1, low = 0):

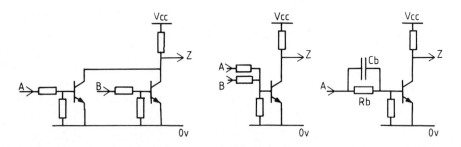

Fig. 2.1 Resistor transistor logic. (a) RTL NOR gate. (b) RTL NOR gate. (c) RCTL.

B	A	Z
0	0	1
0	1	0
1	0	0
1	1	0

This is the truth table for a NOR gate, so the basic RTL element is a NOR gate.

RTL uses a saturating output transistor, so the 0 state is well defined. RTL does, however, have problems in the 1 state. Each input draws base current through the collector load resistor. The greater the number of inputs driven, the lower the 1 voltage becomes. Fan-out is therefore poor, typically 4 to 5.

The design trade-off between overdriving the output transistor and charge storage delay was explained in section 1.5.2. A variation of RTL found in industrial families is called RCTL (Resistor Capacitor Transistor Logic) and uses a base capacitor Cb as shown in fig. 2.1c to give a good turn-on without excessive charge storage. Rb is chosen so the transistor just saturates. Cb, however, gives a pulse of current on a high-going edge which drives the transistor hard on for a few nanoseconds. Once on, the charge in Cb has gone, and the transistor is just held on by Rb. Charge storage at turn-off is minimal. The time constant RbCb is critical, and must be chosen with the system operating speed in mind. Fan-out is increased with RCTL, because the value of Rb is higher.

RTL is obsolete and not currently available in IC form except on a maintenance basis. The most popular families were the Fairchild μL and mWμL families and the Texas Instruments Series 51 (the latter using RCTL techniques). These had, respectively, tpd of 12 ns, 40 ns and 130 ns. The only advantage of these over more modern families is the low power consumption (2 mW for a mWμL package), which approaches that of CMOS.

2.3 Diode Transistor Logic (DTL)

The simple logic family outlined in section 1.3 used diodes. DTL is a practical version, and a typical example is shown in fig. 2.2. When both inputs are high, the output transistor turns on. If we call a high voltage a 1 and a low voltage a 0, we have the truth table:

B	A	Z
0	0	1
0	1	1
1	0	1
1	1	0

This is the truth table for a NAND gate, so the basic DTL gate is a NAND gate.

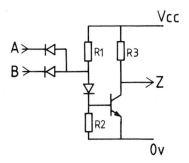

Fig. 2.2 Diode transistor logic.

Inputs in DTL draw no current in the 1 state, but source current in the 0 state. The output voltage in the 0 state is defined by a saturated transistor, which can sink the current from many inputs. Fan-outs in DTL are therefore high (typically 10 to 20) and the logic levels well defined at 0 V and Vcc. Well-defined logic levels give good noise immunity.

DTL can operate quite fast (up to 1 MHz), the main restrictions being charge storage, and stray capacitance degrading the 0 to 1 edge with a time constant R3Cs.

DTL is still available in IC form, and has a reputation for being tough and relatively bullet-proof. Popular DTL families are:

Fairchild	DTμL
Ferranti	Micronor 1 and 2
Motorola	MC200 and MC 250 series
Texas Instruments	Minuteman Series 200
Westinghouse	200, 500 and 800 series

DTL is also widely used in industrial logic families, such as the English Electric Datapac 2.

2.4 Complementary Transistor Logic (CTL)

The ability to manufacture npn and pnp transistors on the same IC allowed an interesting development of DTL. A CTL circuit is shown in fig. 2.3. The input diodes have been replaced by transistors.

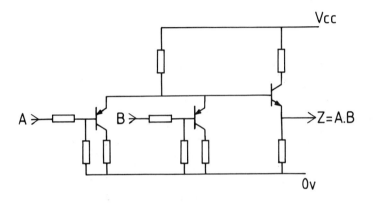

Fig. 2.3 Complementary transistor logic.

Two manufacturers produced CTL; Fairchild CTµL and Texas Instruments Series 53 and 73. Both are no longer available. Interestingly, though, the Series 53/73 is obviously an ancestor of the popular Series 74 TTL. There is some degree of pin compatability between Series 53/73 CTL and Series 54/74 TTL.

2.5 Direct Coupled Transistor Logic (DCTL) and Integrated Injection Logic (I^2L).

DCTL was an early logic family which was never widely used. It has recently reappeared in a revised form as I^2L. A typical DCTL gate is shown in fig. 2.4. Throughout the logic, transistor collector outputs are connected directly to input bases. In fig. 2.4, if either input is positive the output transistor turns off. The circuit is an OR gate.

DCTL has many disadvantages: poor noise immunity, low voltage swing, poorly defined levels. It was consequently not very popular.

I^2L is similar in principle to DCTL, but uses current source transistors instead of resistors, and multiple collector transistors to simplify internal connections. A typical I²L circuit is shown in fig. 2.5. This has two outputs, Z being an OR function and Y an inversion function.

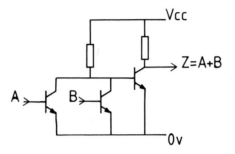

Fig. 2.4 Direct coupled transistor logic (DCTL).

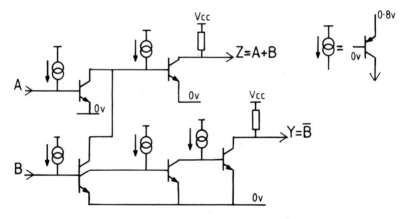

Fig. 2.5 Integrated injection logic (I²L).

$$Z = A + B$$
$$Y = \overline{B}$$

I²L is particularly easy to make in IC form, but suffers from the defects of DCTL. ICs using I²L therefore only use direct coupling within the IC, and use external pull-up resistors to increase the logic swing outside the IC. At the time of writing, I²L is used internally in many MSI and LSI chips, but is not available as a family in the way that DTL, ECL, CMOS or TTL are.

2.6 Emitter Coupled Logic (ECL) and Common Mode Logic (CML)

ECL can claim, quite correctly, to be the fastest commercially available logic family. With propagation delays as low as 1 ns, and operating speeds as high as 500 MHz, ECL is about an order of magnitude faster than its newest rival Schottky TTL. As we shall see in section 11.2, however, speed does bring its own problems.

The two main constraints on speed are charge storage and stray capacitance, as described in section 1.5.2. ECL overcomes these problems by operating with non-saturating transistors (hence no charge storage) and operating at fairly high currents (thereby reducing RC time constants).

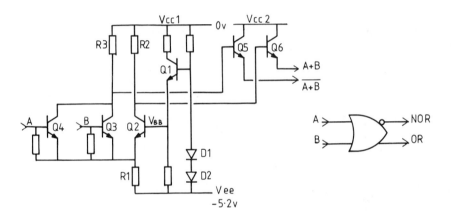

Fig. 2.6 Emitter coupled logic. (a) Circuit. (b) Symbol.

A simple ECL gate is shown in fig. 2.6. At first sight it resembles a DC amplifier long-tailed pair. Transistor Q1, along with diodes D1, D2, generates a bias reference voltage V_{BB} of -1.2 volts which is applied to the base of transistor Q2. The two logic levels in ECL are -0.8 volts and -1.6 volts; 0.4 volts above and below the bias voltage (the noise immunity is consequently rather poor at 0.25 volts). If either input A or B is at -0.8 volts, the majority of the current flowing through R1 comes via R3 and Q3 or Q4. If both A and B are at -1.6 volts, the majority of the current flowing through R1 comes via R2 and Q2. The voltages at Q5 and Q6 bases will be complementary. Transistors Q5 and Q6 act as both level restorers (back to -0.8 volts and -1.6 volt levels) and emitter follower buffers. Low impedance outputs are necessary because the high speeds attained by ECL demand terminated transmission lines, a topic discussed further in section 11.2. These emitter follower outputs give a high fan-out of 25, but this is reduced considerably when terminated lines are used.

The basic gate of fig. 2.6a has two outputs. Q5 provides a NOR function, Q6 provides an OR function. The symbol for the ECL basic gate is shown in fig. 2.6b. The NOR output is denoted by the inversion circle on the output.

On fig. 2.6a it will be seen that there are two Vcc connections. ECL normally runs with a single supply of −5.2 volts connected to Vee, and Vcc connected to ground. The output emitter followers in ECL are designed to drive low impedance terminated transmission lines. This implies high currents with fast turn-on and turn-off times; a very good way of generating power supply noise via supply rail inductance.

Two 0 volt connections are therefore provided on each ECL chip. A 'clean' supply Vcc1 for the logic and a 'noisy' supply Vcc2 for the output emitter followers. These supply rails are run around the printed circuit board on totally separate tracks, only meeting at the power supply.

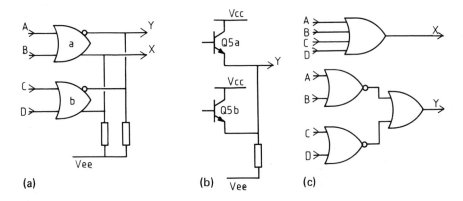

Fig. 2.7 ECL Wire-OR function. (a) Connection. (b) Output stages. (c) Logic diagram.

A great advantage of ECL is the ability to link outputs, thereby generating an additional OR function. In fig. 2.7a we have two linked gates giving outputs X and Y. Two linked outputs are shown in fig. 2.7b. If either transistor goes high the output will be high, giving an OR function between the two gate outputs. The logic diagram for fig. 2.7a is therefore as fig. 2.7c with the two outputs X, Y being:

$$X = A + B + C + D$$
$$Y = \overline{(A + B)} + \overline{(C + D)}$$

This linking of outputs is known as a Wire-OR (sometimes the term Phantom OR is used).

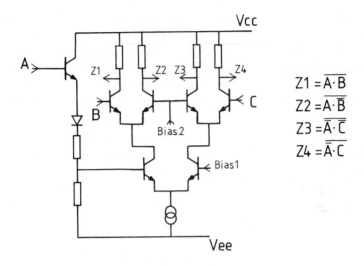

$$Z1 = \overline{A \cdot B}$$
$$Z2 = \overline{A \cdot \overline{B}}$$
$$Z3 = \overline{\overline{A} \cdot \overline{C}}$$
$$Z4 = \overline{\overline{A} \cdot C}$$

Fig. 2.8 ECL series gating.

ECL can provide complex AND/OR combinations by modification of the basic circuit. Further splitting of the source current, as in fig. 2.8, provides the simple NAND function, a technique known as series gating. A combined AND/OR function can be obtained by linking collectors.

The high speed of ECL means that more care needs to be taken in its use; in general ECL can only be used on a printed circuit board and is not really suited for 'one-offs' constructed with wire wrap or soldered wire link boards.

ECL is manufactured mainly by Fairchild and Motorola, who were amongst the originators with their MECL 1 range introduced in the early 1960s. Their current range is typical of most manufacturers and consists of the following ECL families:

	Propagation delay (ns)	Edge speed (ns)	Operating speed (MHz)	Power (mW)
MECL 10K	2	3	200	25
MECL 10KH	1	1.5	250	25
MECL 111	1	1	500	60

34

These all use a −5.2 volt rail and can be intermingled, and there is a large degree of pin compatability between the various families. There is little compatability between Motorola and Fairchild ECL.

MECL 10KH is presented as a 'general-purpose' ECL family, giving a speed which is more than adequate for most applications, and having easier connection rules than its faster MECL 111 cousin. Power supply tolerance is also improved, as MECL 10KH uses a constant current source in place of R1 on fig. 2.6a, and an improved V_{BB} generator which tracks with supply variation.

ECL, therefore, is a very fast logic family. This high speed requires quite complex wiring rules to be followed (described further in section 11.3). For most applications the high speed is unnecessary, and TTL or CMOS, described below, is simpler and cheaper to use. Where the ultimate in speed is required, however, the designer has little choice other than ECL.

2.7 Transistor Transistor Logic (TTL)

TTL is probably the most popular and most successful logic family. It is a direct descendant of DTL, with many substantial improvements. A TTL NAND gate is shown in fig. 2.9. The rather odd-looking double emitter transistor acts like the input diodes on the DTL gate of fig. 2.2. TTL operates on a 5 volt supply with a tolerance of ± 0.25 V, using a 1 level of 3.5 V and a 0 level of 0 V.

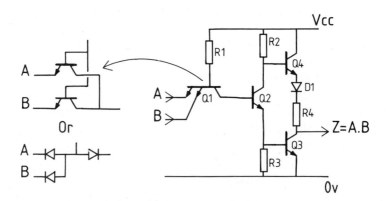

Fig. 2.9 Transistor-transistor logic (TTL).

If both inputs AB are high, base current will flow into Q2 via R1. Q2 turns on, turning Q3 on in turn, which gives an output voltage of 0 V.

If either, or both, inputs are low, current flows from R1 to the gate(s) driving the inputs. Q2 is turned off and Q4 acts as an emitter follower, taking the output to about 3.5 volts. The gate thus has the truth table:

B	A	Z
0	0	1
0	1	1
1	0	1
1	1	0

which is the truth table for a NAND gate.

The output transistors Q3, Q4 are known as a totem pole output, and have several advantages over the simple pull-up resistor of fig. 2.2.

With the DTL gate of fig. 2.2, R3 has to be kept low to reduce RC time constants with stray and line capacitance. When the gate output is 0, Q1 is turned on and R1 passes a relatively high current. The gate power consumption is therefore significantly higher in the 0 output state than in the 1 output state which can cause power supply noise. The totem pole arrangement gives substantially the same power consumption in both states.

In the 0 state, Q3 is a saturated transistor; in the 1 state Q4 is an emitter follower. The gate output impedance is therefore low in both states, which gives short RC time constants and well-defined output levels.

The totem pole output does, however, have one disadvantage. As the output changes state, both Q3 and Q4 can come on together for a few ns. This causes a high current pulse to be taken from the supply, which can be as high as 50-100 mA. This pulse has fast edges which can, with supply lead inductance, cause noise spikes on the 5 volt and 0 volt rails. TTL normally requires decoupling capacitors at regular intervals on the logic board. The recommended decoupling is one 0.01 μF capacitor per five IC packages, but most designers use one 0.01 μF per IC package. This is not as complex or as costly as may be first thought. Decoupling of TTL and supply noise problems are discussed further in chapter 13.

When TTL is driven from a source other than another TTL gate it is important to appreciate the current requirements of a TTL input. When an input is low, as in fig. 2.10a, the driving output is required to *sink* current via R1. This current is 1.6 mA in normal TTL. When an input is high as in fig. 2.10b, the driving output is required to *source* leakage current into Q1. This current is typically 40 μA. This sourcing of current in the high state and sinking of current in the low state is normally provided by the totem pole output.

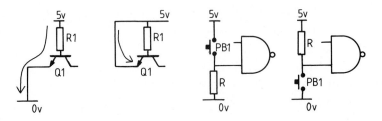

Fig. 2.10 Input requirements of TTL. (a) Input low. (b) Input high. (c) Pull down. (d) Pull up.

If a gate input is left unconnected, Q2 will be turned on with current from R1. A floating input therefore looks like a 1. It is not recommended that unused inputs be left floating, however, as they are rather prone to noise.

TTL logic levels are nominally 3.5 volts for a 1 and 0 volts for a 0. The 1 level can fall to 2.4 volts, and the 0 level can rise to 0.4 volts. The threshold voltages are 2.0 volts and 0.8 volts. TTL therefore has a noise immunity of 0.4 volts minimum in both states. This is a worst case value and a typical gate would have a noise immunity over 1 volt.

Knowledge of TTL input levels and currents is necessary if it is desired to interface TTL to other logic families or discrete components. In fig. 2.10c, for example, a gate input is tied to 0 volts via a resistor R, forcing a 0 into a gate input. When PB1 is pressed the input goes to a 1 as the input goes to 5 volts. We know that the maximum 0 level is 0.4 volts, and the input supplies 1.6 mA in the 0 state. The maximum value of R is given by:

$$R = \frac{0.4 \text{ V}}{1.6 \text{ mA}} = 250 \text{ ohms}$$

220 ohms is the next lowest preferred value. Similar calculations give 390 ohms for LS TTL, described later.

Similarly, in fig. 2.10d a gate input is pulled up to a logic 1 level by R, and switched down to 0 volts by PB1. The maximum value of R is given by:

$$R = \frac{(5 - 3.5)}{40\mu A} \text{ volts} = 37.5 \text{ K ohms}$$

33 K ohms is the next lowest preferred value, although 10K is normally used.

The current sourcing/sinking requirements of TTL provide particular difficulties when mixing TTL and CMOS as we shall see in section 13.4.4.

The circuit of fig. 2.9 is known as standard, or normal, TTL. It has a propagation delay of 10 ns, and a power consumption of 10 mW per gate. There are many variations on standard TTL. By increasing the resistance values, a low power TTL is produced. This has a power consumption of 1 mW per gate, but the propagation delay increases to 33 ns.

By reducing the resistance values, a high-speed TTL can be produced. This has a propagation delay of 6 ns, but a power consumption of 22 mW.

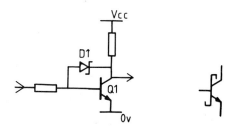

Fig. 2.11 Schottky clamping.

Standard, low power and high speed TTL are all saturating logic, and suffer from charge storage delays. In fig. 2.11 we have a transistor Q1 with a Schottky diode connected from base to collector. A Schottky diode has a very low forward voltage drop, typically 0.3 volts. A saturated transistor has a V_{be} of 0.8 V and a V_{ce} of 0.2 volts. As the transistor starts to saturate, D1 will conduct, limiting V_{ce} to about 0.5 volts. The transistor cannot saturate, and consequently does not suffer from charge storage delay. The arrangement of fig. 2.11 is known as Schottky clamping, and gives the fast turn-on of a saturated transistor without the turn-off delay.

Schottky TTL uses Schottky-clamped transistors to achieve an increase in speed (propagation delay 3 ns) without a dramatic increase in power (power consumption 19 mW).

Low power Schottky increases the resistance values to give a performance similar to standard TTL (propagation delay 9.5 ns) at a much lower power (power consumption 2 mW). At the time of writing low power Schottky is the most versatile and widely used TTL variant.

A recent addition to the TTL range is Advanced Schottky. This uses improved integrated circuit construction techniques allowing smaller geometry on the IC 'slice', giving higher speeds for no power increase. This is available in ALS form (for Advanced Low Power Schottky) having a propagation delay of 4 ns and a power consumption of 1 mW.

The six versions of TTL are:

Name	Suffix	Propagation delay (ns)	Power/gate (mW)
Normal (Standard)	None or N	10	10
Low power	L	33	1
High speed	H	6	22
Schottky	S	3	19
Low power Schottky	LS	9.5	2
Advanced low power Schottky	ALS	5	2

The suffix appears as a part of the device numbering. 74LS06, for example, is a low power Schottky device.

All the TTL variants have common pin connections, and can be intermixed freely provided the fan-out and fan-in restrictions of each family are considered. If we designate the input of a Standard TTL gate as one unit load, we can build up a table in terms of unit loads:

Family	Suffix	Fan-out (UL)	Fan-in (UL)
Normal	None or N	10	1
Low power	L	2.5	0.25
High speed	H	12.5	1.25
Schottky	S	12.5	1.25
Low power Schottky	LS	5	0.5
Advanced low power Schottky	ALS	5	0.5

From the table, it can be seen that, for example, a normal TTL gate can drive 50 LS inputs, or four Schottky inputs and five normal inputs. Different TTL families can be freely mixed provided fan-out loadings are not violated. Note that within a family, a gate output can always drive 10 gate inputs.

TTL first appeared in the early 1960s in two versions. Texas Instruments introduced its TTL range with its 74 series (industrial 0-70°C) and 54 series (full military specification −55°C to +125°C plus wider supply tolerance). Sylvania introduced its SUHL range (Sylvania Universal High Speed Logic). Electrically the 74/54 series and SUHL were similar, but the pin connections were totally different. In particular, 74/54 series put the supply connections on the diagonal corners (pin 7 for 0 V and pin 14 for 5 V on a 14-pin package) whereas SUHL had them in the middle (pin 10 for 0 V and pin 4 for 5 V on a 14-pin package). At the time there was considerable rivalry between Texas Instruments and Sylvania over the relative merits of the 74/54 series and SUHL, which the market decided in favour of Texas Instruments. Today, SUHL and its second source suppliers such as Transitron, are only available on a maintenance basis. Virtually all TTL is now pin-compatible and electrically compatible with 74 series TTL.

2.8 Complementary Metal Oxide Semiconductor (CMOS) logic

CMOS is virtually the perfect logic family. It can operate on a wide range of power supplies (from 3 to 15 volts), uses little power at low speeds (typically 0.01 mW), has high noise immunity (approximately 30% of the supply voltage) and simple interconnection rules due to its high fan-out (typically in excess of 50). Its one slight disadvantage is a maximum operating speed of about 5 MHz; somewhat slower than TTL and considerably slower than ECL.

CMOS is built around the two types of field effect transistor (FET) shown in fig. 2.12. The construction of these two types of FET need not concern the user. The n channel FET of fig. 2.12a can be considered as a switch in series with a 1K resistor. The switch is controlled by the voltage between the gate and source. When the gate is positive with respect to the source the switch is closed; when V_{GS} is zero the switch is open. Note that the n channel FET operates with the drain connection positive with respect to the source.

The p channel of fig. 2.12b operates with the drain connection negative with respect to the source, and can again be considered as a simple switch. When the gate is negative with respect to the source, the switch is closed; when V_{GS} is zero the switch is open.

It is possible to construct a logic family solely from n channel or p channel FETs, but virtually all MOS logic uses both (hence the name CMOS). A typical CMOS inverter is shown in fig. 2.13. When the input is at a 1, Q1 is turned off and Q2 is turned on (by the rules of fig. 2.12). The output is thus connected to 0 V by Q2 which appears as a 1K resistor. When the input is a 0, Q1 is turned on and Q2 is turned off. The output is now connected to the positive supply rail via Q1, which again appears as a 1K resistor.

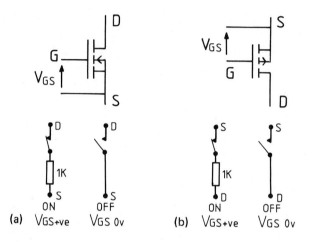

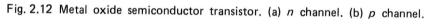

Fig. 2.12 Metal oxide semiconductor transistor. (a) *n* channel. (b) *p* channel.

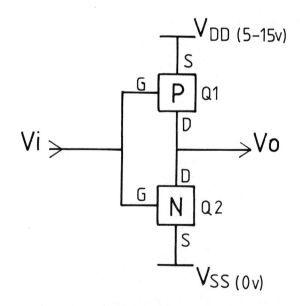

Fig. 2.13 CMOS inverter.

CMOS outputs with no load have virtually no offset voltage from V_{DD} or 0 V, but are output-current dependent. A load drawing, say, 1 mA will degrade the output by 1 volt because of the inherent 1K on-resistance of the *n* and *p* channel FETs. This effect is particularly important when mixing CMOS with other logic families or with discrete components.

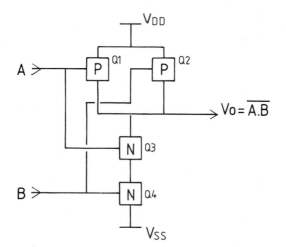

Fig. 2.14 CMOS NAND gate.

It is a simple matter to construct other logic gates with CMOS. In fig. 2.14, the output will only be switched to 0 V if both A and B are 1, thereby turning both Q1, Q2 off and Q3, Q4 on. Figure 2.14 is therefore a NAND gate.

In fig. 2.15, if either A or B is a 1, Q1 or Q2 will turn off and Q3 or Q4 will turn on, taking the output to 0 V. Figure 2.15 is therefore a NOR gate.

Obviously, multi-input gates can be constructed, and by combining the NAND and NOR gates with inverters the AND and OR functions can be obtained. One point of note is that CMOS gates use no components other than the *n* channel and *p* channel FETs. This simplifies the design and construction of CMOS integrated circuits and leads to lower costs.

A CMOS input is simply the gate of a FET, and has almost an infinite resistance. Fan-out is accordingly very high. This is not, however, the full story. A CMOS output appears as a 1K resistor connected to the positive supply (for a 1) or 0 V (for a 0). During switching, a CMOS output will follow an RC exponential curve where R is 1K and C is the combination of gate input capacitance (typically 5 pF per input) and connecting line capacitance (typically 30 pF per metre). Fan-out is therefore determined by the maximum rise and fall times that can be tolerated. In almost all systems a fan-out of 50 can be used without problems.

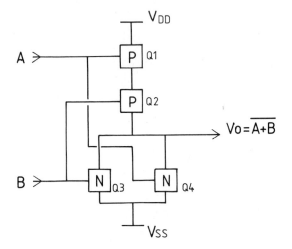

Fig. 2.15 CMOS NOR gate.

The high input impedance of a FET can present handling problems. Static electricity, generated by nylon clothing for example, can cause potentials of many thousands of volts. This high voltage cannot leak away with the high impedances present in CMOS, and in extreme cases can cause device failure. Early CMOS was rather prone to damage in this way, and careful handling rules were recommended, including assembly on metal foil, earthed wrist bands and controlled humidity areas. Later CMOS now includes protection diodes as shown in fig. 2.16 and can be handled almost like any other component. A few simple handling rules should still be followed to ensure 100 per cent protection:

(a) Do not remove CMOS chips from their conductive foam until just before they are needed.
(b) Do not leave CMOS chips lying around out of their foam.
(c) Always insert CMOS chips last when constructing a logic board so that gate inputs have a route to ground.
(d) On multi-board systems, if a gate input comes onto a board via an edge connector, use a 100K resistor from the input to ground

so the board can be handled without fear of static damage.
(e) Always use an earthed soldering iron.

Another effect of the high input impedance is the tendency of
unused inputs to charge to an unpredictable voltage. *All* CMOS
inputs must go somewhere; unused inputs, even on unused gates,
must be tied to either the supply or 0 V dependent on the function.
In extreme cases unused inputs can charge to a voltage sufficiently
high to cause device failure during handling, so the designer must
always carefully check for unused inputs.

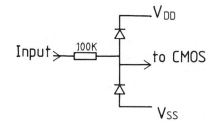

Fig. 2.16 CMOS input protection.

CMOS originally appeared in the forms shown in figs 2.13 to 2.15,
known as A series. The later B series incorporates a buffered output,
improved static protection, wider supply tolerance and slightly faster
speed. B series should be used for all new designs.

The main attraction of CMOS is its simple power supply requirements. CMOS can run on any voltage between 3 and 15 volts (18
volts for B series). The higher voltages give faster speeds, and the
'optimum' voltage is about 12 volts for A series and 15 volts for B
series. This gives maximum speed with a margin for supply overvoltage.

Although CMOS has minimal power requirements, as the operating
speed rises the charge and discharge currents of stray capacitances
causes the operating current to rise considerably. At low frequencies
a CMOS gate requires about 10 nW. Above 500 kHz the power consumption rises to about 0.5 mW which is similar to LS TTL operating
at about the same speed. At the limit of CMOS operating speed
(about 5 MHz) LS TTL actually uses slightly less power. Overall
power consumption of a system does not go up in a similar way, however, because all the gates do not switch at once. In a practical large
system operating at about 1 MHz, CMOS will still have almost one-
tenth of the power requirements of an equivalent LS TTL circuit.

There is almost universal pin compatibility between different CMOS manufacturers with the so-called 4000 series. This is a rational combination of the original RCA COSMOS and Motorola McMOS (manufactured in Scotland, hence the name!). The 4000 series in A and B versions is available from almost all manufacturers.

A useful CMOS variant is the so called 74C series. This is a CMOS family which is pin compatible (but *not* electrically compatible) with 74 series TTL. It is mainly intended for producing low power versions of existing TTL circuits; the 4000B is better suited to totally new designs. All 74C devices have B series characteristics.

A recent MOS development is SOS, for Silicon on Sapphire. This uses sapphire as an insulating material, and reduces the speed limiting stray capacitances. SOS can operate ten times as fast as CMOS, but at the time of writing is prohibitively expensive for most applications. It does hold great promise for high-speed, low-power circuits when its cost inevitably falls.

2.9 Industrial logic families

An often overlooked area of logic design is concerned with industrial control. Industrial logic operates at relatively low speed and works with limit switches, hydraulic valves, motors and similar items of heavy machinery.

Industrial logic is designed to be virtually indestructible, and is usually a variation on RTL or DTL. Frequently discrete components are encapsulated in epoxy resin, giving higher voltage ratings than are available with ICs. The characteristics of industrial logic are:

(a) Virtually indestructible; ability to stand 110 volt AC on inputs.
(b) Slow speed; 100 kHz typical maximum speed.
(c) Wide supply range, typically 5 V to 30 V.
(d) Physically large, often using Lucar connectors for wiring.
(e) High noise immunity.

A typical industrial logic family is the Mullard NORBIT range, which is a complete family of gates, flip-flops, monostables, etc., using RTL. Other popular industrial logic is NORLOG, Datapac, Sigmatronic and SSR.

There have also been high noise immunity ICs manufactured for use in noisy applications, but these have been largely superseded by CMOS. A typical example is the HTL gate shown in fig. 2.17. This is a DTL gate operating on a 15 volt supply with the zener diode giving increased noise immunity.

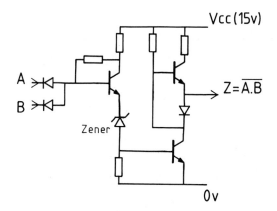

Fig. 2.17 High noise immunity NAND.

2.10 Future developments

The LS variant of TTL and the B Series CMOS have become established as industry standards. Technology does not, however, stand still and many variations on CMOS and TTL are appearing.

A very promising development is the High Speed CMOS available from Motorola and other suppliers. This family is deliberately designed to combine the speed of LS TTL (8 μs propagation delays) with CMOS power consumption (1 μW static consumption per gate, although static power consumption is not the whole story as explained in section 1.5.2). Rather interestingly, the family has been designed with 74 series pinning rather than 4000 Series CMOS pinning, and for a supply of 2 to 6 V. With a fan-out of 8 Standard LS TTL loads, it is evidently aimed as a TTL chip for chip replacement.

There are interesting variations on Schottky TTL (such as the Fairchild FAST family) but it is not really technical excellance that establishes a logic family. Above all else, price, availability, range of devices and multi source stockists make a logic family successful. It is the standardisation of TTL and CMOS that has established their position. Any future developments will probably relate to 74 Series or 4000 series (as the Motorola High Speed CMOS does).

CHAPTER 3
Combinational Logic

3.1 Introduction

The simplest logic systems are those which do not involve counters or storage elements. Such systems have several inputs and one or more outputs, and the output state(s) are uniquely defined for every combination of inputs. These logic systems are known as 'combinational logic' or 'static logic' systems and are constructed solely from logic gates.

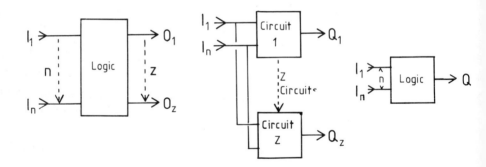

Fig. 3.1 Combinational logic block diagrams. (a) Generalised problem. (b) Separate circuits. (c) One circuit.

A combinational logic system can therefore be represented by fig. 3.1a. We have n inputs I_1 to I_n and z outputs O_1 to O_z. Between them there is a circuit of logic gates such that the outputs relate to the inputs in the required manner. In systems with multiple outputs it is usually more convenient (and simpler) to consider each output separately as in fig. 3.1b. A combinational logic system can therefore be constructed as a collection of circuits similar to fig. 3.1c where we have n inputs and one output. The state of the output is determined solely by the state of the inputs.

The number of possible input states depends on the number of inputs:

For two inputs there are four possible input combinations
For three inputs there are eight possible input combinations
For four inputs there are sixteen possible input combinations
and so on.

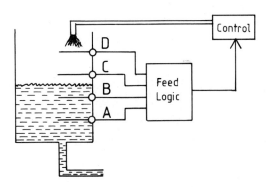

Fig. 3.2 System not using 'ALL' input states.

Not all the possible input conditions might be used, however. In fig. 3.2 we have a storage tank with four level switches ABCD as inputs to a combinational logic circuit. For four inputs there are 16 possible combinations. Except for fault conditions, the designer of this system need only consider five input states:

D	C	B	A
0	0	0	0
0	0	0	1
0	0	1	1
0	1	1	1
1	1	1	1

It should be noted, though, that a combinational logic will give 'some' output for any input state, and the cautious designer would check that nothing untoward occurs if, say, the input state A = 1, B = 0, C = 1, D = 1 appeared due to a switch failure.

This chapter is concerned with the design of combinational logic circuits. Very few real-life systems are purely static by nature; most involve storage, feedback or counters. These real-life systems can, however, be analysed and designed by considering them as smaller subsystems connected together. The techniques outlined in this chapter provide the basis for constructing combinational subsystems which are in turn used to make computers, industrial controllers, calculators, and almost every digital device.

3.2 Logic gates

3.2.1 Introduction
Logic gates were introduced briefly in chapter 1. In this section the simple gates are reviewed, and dealt with in more detail. The use of these gates as the basic building blocks of combinational logic is described.

3.2.2 The Non-inverting buffer and the inverter
The simplest logic gates are those with one input. There are two single-input gates, the non-inverting buffer and the inverter. The buffer is shown in fig. 3.3a. The output Z follows the input, or in Boolean algebra:

$$Z = A$$

At first sight the buffer is a rather superfluous item, but it is usually employed to overcome fan-out problems. Buffers with high fan-outs are available in all logic families.

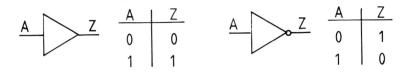

Fig. 3.3 Single-input gates. (a) Non-inverting buffer. (b) Inverter.

The inverter is used to complement (or invert) an input. The small circle on the output of fig. 3.3b denotes inversion. In Boolean algebra:

$$Z = \overline{A}$$

Buffers and inverters are usually found in hex packages (i.e. six gates to one IC) such as the 7406 hex inverter in TTL or the 4050 hex buffer in CMOS. Frequently, buffers and inverters incorporate tri-state outputs, described further in section 3.2.12.

3.2.3 AND gates

The output of an AND gate is a 1 if, and only if, all its inputs are 1, otherwise its output is 0. The four-input AND gate of fig. 3.4a will have $Z = 1$ only if A, B, C, D are all 1. In Boolean algebra:

$Z = A.B.C.D$

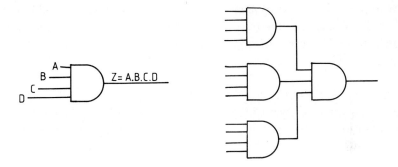

Fig. 3.4 AND gates. (a) Four-input AND. (b) Twelve-input AND.

AND gates are commonly available in 2, 3 and 4-input forms. Any desired number of inputs can be obtained by cascading gates; a twelve-input combination is shown in fig. 3.4b. Unused AND gate inputs should be paralleled with a used input or tied to a 1 level.

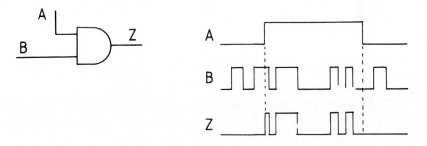

Fig. 3.5 Gating a signal with an AND gate. (a) Circuit. (b) Waveforms.

An AND gate can literally be used to gate a signal. In fig. 3.5 we have a signal occurring on input B of an AND gate. When input A is 1, the output Z follows B. When A is 0, output Z is 0. Input A can be said to gate signal B.

AND gates are available with open collector outputs and tristate outputs, described further in section 3.2.12.

3.2.4 OR gates

The output of an OR gate is 1 if *any* of its inputs is 1. The four-input OR gate of fig. 3.6a will have Z = 1 if A or B or C or D is 1. In Boolean algebra:

$$Z = A + B + C + D$$

OR gates are commonly available in 2, 3 and 4-input forms, but any number of inputs can be obtained by cascading gates; a six-input combination is shown in fig. 3.6b. Unused OR gate inputs should be paralleled with a used input or tied to a 0 level.

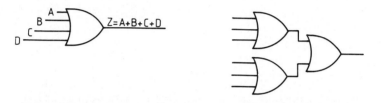

Fig. 3.6 The OR gate. (a) Four-input OR. (b) Six-input OR.

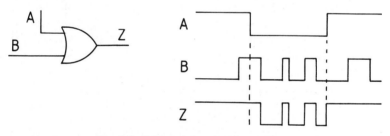

Fig. 3.7 Gating a signal with an OR gate.

An OR gate can also be used to gate a signal in a similar way to an AND gate. In fig. 3.7 the OR gate passes a signal when input A is 0, and blocks (output Z a permanent 1) when A is 1.

3.2.5 NAND gates

The NAND gate is the most commonly used logic gate for reasons we shall explain in section 3.2.7. (The 7400 Quad two-input TTL NAND is probably the worlds most used IC.) The NAND gate is simply an AND gate followed by an inverter, as in fig. 3.8a, so the output Z is 0 if, and only if, both inputs AB are 1; otherwise Z is 1. In Boolean algebra:

$$Z = \overline{A.B}$$

The symbol for a NAND gate is shown in fig. 3.8b, the small circle at the output denoting the inversion.

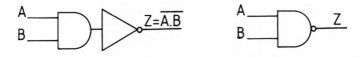

Fig. 3.8 The NAND gate. (a) Function. (b) Symbol.

NAND gates are available in 2, 3, 4, 8 and 13-input forms. Early TTL had input expander connections on some gates allowing virtually any desired NAND gates to be built, but the expander inputs were rather noise-prone and have generally been dropped from current ranges.

Because of the inverted outputs, NAND gates cannot be cascaded in a similar way to AND and OR gates, but the 13-input gate caters for most applications. Unused inputs should be paralleled with a used input or tied to a 1 level.

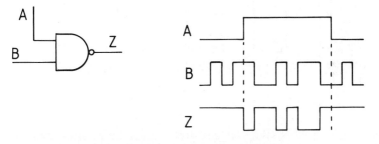

Fig. 3.9 Gating a signal with a NAND gate.

When a NAND gate is used to pass a signal, the output is an inverted version of the input as shown in fig. 3.9, with the blocked state being a 1.

3.2.6 NOR gates

The NOR gate is almost as common as the NAND gate for reasons explained in the next section. The NOR gate is an OR gate followed by an inverter as shown in fig. 3.10a, so the output Z is 0 if *any* of the inputs AB is 1, otherwise Z is 1. In Boolean algebra:

$$Z = \overline{A + B}$$

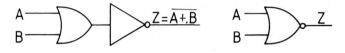

Fig. 3.10 The NOR gate. (a) Function. (b) Symbol.

The symbol for a NOR gate is shown in fig. 3.10b, the small circle at the output again denoting the inversion.

NOR gates are available for 2, 3, 4 and 8-inputs. Additional inputs cannot be obtained by cascading gates because of the inverted output. Unused inputs should be paralleled to a used input or tied to a 0 level.

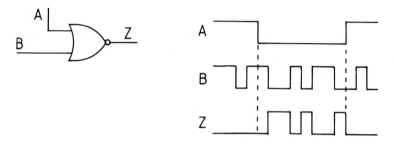

Fig. 3.11 Gating a signal with a NOR gate.

If a NOR gate is used to pass a signal as in fig. 3.11a the gated signal is an inverted form of the input with the blocked state being a 0.

3.2.7 Universal gates

The NAND and NOR gates are sometimes called universal gates. It can be shown that any required logic circuit can be constructed solely with NAND gates or NOR gates. This has obvious attractions in that the designer can standardise on relatively few ICs.

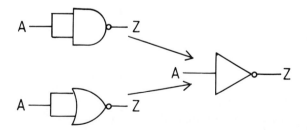

Fig. 3.12 Use of NAND/NOR gates as inverters.

By paralleling inputs as in fig. 3.12, NAND gates and NOR gates can be used as inverters.

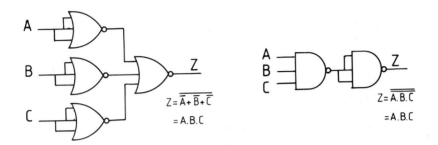

Fig. 3.13 Implementation of an AND gate. (a) NOR circuit. (b) NAND circuit.

An AND gate can be constructed from NAND gates or NOR gates as shown in fig. 3.13. The NOR gate circuit is a direct result of De Morgan's theorem since:

$$Z = \overline{\overline{A} + \overline{B} + \overline{C}} = A.B.C$$

An OR gate can be also constructed from NAND gates or NOR gates as shown in fig. 3.14. The NAND gate circuit is again derived from De Morgan's theorem since:

$$Z = \overline{\overline{A}.\overline{B}.\overline{C}} = A + B + C$$

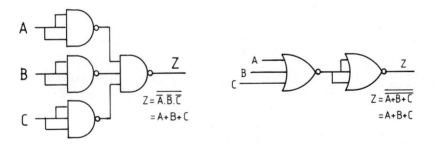

Fig. 3.14 Implementation of an OR gate. (a) NAND circuit. (b) NOR circuit.

A designer usually works solely with NAND gates or NOR gates. The choice is determined by gate availability in the family being used and the designer's experience. TTL is the most widely used family, and is rather NAND orientated (NAND gates are cheaper and available in a wider range of input configurations than NOR gates). Most logic designs are therefore done in NAND logic rather than NOR logic, but the reasons are historical rather than theoretical.

It might be thought that the arrangements of figs 3.12 to 3.14 are rather cumbersome and wasteful of gates. In practice, it is usually found that inverters cancel in most circuits.

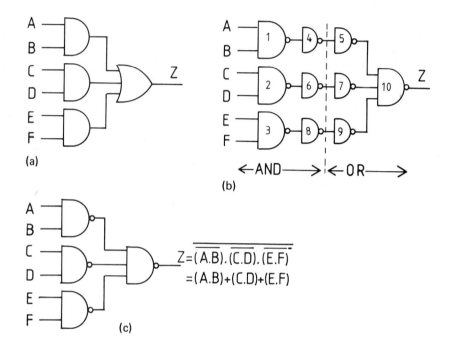

(a)

(b)

$\leftarrow$ AND $\longrightarrow$ $\leftarrow$ OR $\longrightarrow$

$$Z = \overline{\overline{(A.B)} . \overline{(C.D)} . \overline{(E.F)}}$$
$$= (A.B) + (C.D) + (E.F)$$

(c)

Fig. 3.15 AND/OR circuit. (a) Required function. (b) Direct implementation. (c) Simplified circuit.

In section 1.4.3 it was shown that a Boolean expression can be expressed in S of P form (Sum of Products) or P of S form (Product of Sum). This will be discussed further in section 3.3. Consider, however, the S of P expression:

$$Z = (A.B) + (C.D) + (E.F)$$

In straight logic form this would be constructed as in fig. 3.15a. Using the circuits of figs 3.12 to 3.14 we construct the rather horrendous-looking circuit of fig. 3.15b. From simple observation, gates 4 to 9 are redundant. Gate 4, for example, inverts the output of gate 1, but gate 5 inverts it back again! Figure 3.15b can thus be simplified to fig. 3.15c. This is, in fact, yet another result of De Morgan's theorem, since:

$$Z = \overline{(\overline{A.B}).(\overline{C.D}).(\overline{E.F})} = (A.B) + (C.D) + (E.F)$$

NOR gates can be used to construct P of S circuits. The expression:

$$Z = (A + B).(C + D)$$

can be generated by the circuit of fig. 3.16 in a similar manner to fig. 3.15.

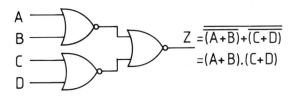

Fig. 3.16 OR/AND circuit.

Almost all combinational logic can be expressed conveniently in S of P or P of S form, so NAND-only logic or NOR-only logic is a useful way of constructing logic circuits from Boolean expressions or truth tables.

3.2.8 Positive and negative logic
Manufacturers' data sheets specify gate operations with a high voltage being a 1 and a low voltage being a 0. TTL, for example, uses 3.5 V for a 1 and 0 V for a 0. This is known, for obvious reasons, as positive logic. If we reverse the convention and call a high voltage a 0 and a low voltage a 1 we are using negative logic.

Consider a positive logic TTL NAND gate. This has the familiar truth table:

B	A	Z
0	0	1
0	1	1
1	0	1
1	1	0

In voltage terms this is:

B	A	Z
0	0	3.5
0	3.5	3.5
3.5	0	3.5
3.5	3.5	0

If we use the same gate, but adopt a negative logic convention (i.e. 3.5 V = 0, 0 V = 1) we get the truth table:

B	A	Z
1	1	0
1	0	0
0	1	0
0	0	1

This is the truth table for a NOR gate. The same gate can be considered as a positive logic NAND gate or a negative logic NOR gate. This is known as 'duality'.

Duality at first may seem confusing. The gate operation does not change, it is the description that changes according to the convention. In English we can describe the TTL NAND gate in two ways:

(a) The output is low if both inputs A and B are high, otherwise the output is high.

(b) The output is high if either input A or B is low, otherwise the output is low.

Description (a) is using positive logic convention; description (b) is using negative logic convention. Both are equally valid.

All gates exhibit duality. The relationships in fig. 3.17 can be proved by substituting 1 for 0 and 0 for 1 in the truth tables. Gates using negative logic use the small circles on inputs or outputs to denote the logic convention. There is no inherent contradiction between this and the circles on the outputs of NAND/NOR gates, as the latter are really AND/OR gates using negative logic convention at the output.

Duality is another way of considering the circuits of figs 3.15 and 3.16. In fig. 3.15c, for example, we are really using negative logic conventions at the input to gate 4, making gate 4 a negative logic NOR gate. Figure 3.15c could be redrawn as fig. 3.18a, although in practice this is rarely done. In a similar way, fig. 3.16c could be redrawn as fig. 3.18b.

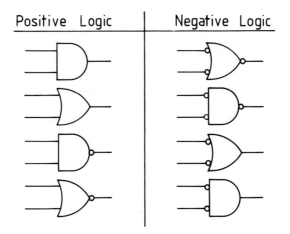

Fig. 3.17 Duality of logic gates.

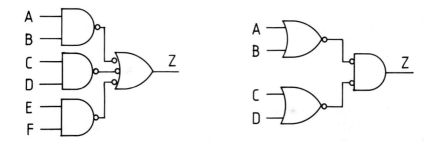

Fig. 3.18 Representations using duality. (a) Function of fig. 3.15. (b) Function of fig. 3.16.

3.2.9 Exclusive OR gates

The exclusive OR gate can only exist in two-input form with the symbol of fig. 3.19a. The output is 1 when its two inputs are different, and 0 when the two inputs are the same:

B	A	Z
0	0	0
0	1	1
1	0	1
1	1	0

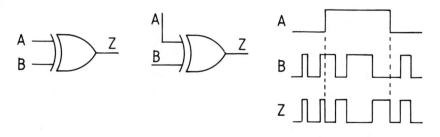

Fig. 3.19 The Exclusive OR (XOR) gate. (a) Symbol. (b) Controlled inverter circuit. (c) Operation.

This is similar to the conventional OR gate except that it excludes the case A = B = 1 (hence the name).

The exclusive OR function (usually written XOR) is not specifically recognised in classical Boolean algebra, but is very useful in constructing logic circuits where digital signals are used to represent numbers (a topic discussed at length in chapters 6 and 8). The exclusive OR function is usually indicated by the pseudo-Boolean symbol $\oplus$. Figure 3.19a can be represented:

$$Z = A \oplus B$$

Examination of the truth table shows that if B is considered as a gating signal, and A as the signal, the XOR behaves as a controlled inverter.

If A = 0 then Z = B
If A = 1 then Z = $\overline{B}$

Similarly

If B = 0 then Z = A
If B = 1 then Z = $\overline{A}$

This action is summarised in figs 3.19b and c.

The XOR gate can only exist in two-input form. If XOR gates are cascaded as in fig. 3.20, the resulting circuit is called a parity generator. The output Z is 1 if the number of inputs ABCD at 1 is odd; Z is 0 if the number of inputs that are 1 is even (e.g. ABCD = 0101 gives Z = 0; ABCD = 1011 gives Z = 1). This is known as a parity generator circuit, a topic discussed further in section 11.5.4.

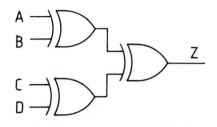

Fig. 3.20 Parity generator (Parity tree).

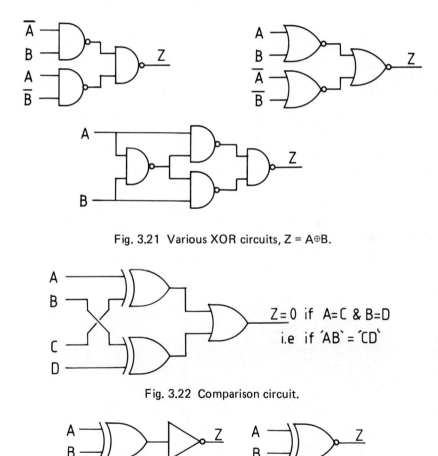

Fig. 3.21 Various XOR circuits, Z = A⊕B.

Z = 0 if A=C & B=D
i.e if ´AB` = ´CD`

Fig. 3.22 Comparison circuit.

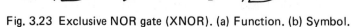

Fig. 3.23 Exclusive NOR gate (XNOR). (a) Function. (b) Symbol.

Although XOR gates are available (e.g. 7486 in TTL; 4030, 4070 in CMOS) it is possible to construct XOR gates with conventional

NOR gates. Figure 3.21 shows various ways this can be done. These are all equally correct; in any particular application the choice of a ready-made gate or one of the circuits in fig. 3.21 will depend on the family being used and the degree of gate standardisation required.

If XOR gates are cascaded with an OR gate as in fig. 3.22, the circuit tests that the input pattern AB is the same as the pattern CD. If the two are equal, Z is 0; if they are unequal Z is 1. This circuit can be expanded to compare any number of inputs.

Figure 3.23a has an XOR gate followed by an inverter. This is known as an exclusive NOR gate, and has the symbol shown in fig. 3.23b. The truth table for an exclusive NOR gate is:

B	A	Z
0	0	1
0	1	0
1	0	0
1	1	1

The exclusive NOR gate, for obvious reasons, is known as a comparison gate. It can be used as an equality tester in a similar manner to XOR gates.

3.2.10 Miscellaneous gates
The commonest logic gates are the inverter, buffer, OR, AND, NAND, NOR, XOR, XNOR. Other gates and gate combinations may occasionally be encountered.

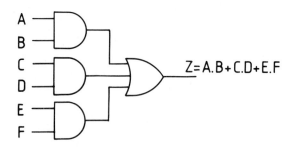

Fig. 3.24 AND/OR (S of P) circuit.

As we shall see in section 3.3, it is convenient to design combinational logic in Sum of Products (S of P) form. This implies an AND-

OR combination of gates similar to fig. 3.24 which has the Boolean equation:

$$Z = A.B + C.D + E.F$$

The AND-OR combination and AND-OR-INVERT combination are available in IC form and are often an economic way of implementing an S of P expression. A typical AND/OR combination is the TTL 7454 which provides four two-input AND gates feeding an OR gate.

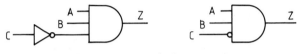

Fig. 3.25 AND/NOT gate.

Industrial logic frequently accepts inputs from limit switches To avoid inverters, industrial logic sometimes uses AND/NOT gates or OR/NOT gates. These have inverters on one (or more) inputs. An AND/NOT gate and its logic symbol is shown in fig. 3.25. Its Boolean equation is:

$$Z = A.B.\overline{C}$$

3.2.11 Schmitt triggers

Many logic elements (flip-flops, monostables, etc., described later) require fast rising and falling edges for reliable operation. Edges can be degraded for a variety of reasons; stray capacitance or even a signal from some slow external device. The Schmitt trigger always gives fast edges on its output signal regardless of the input edge speed.

The transfer function of a conventional gate is shown in fig. 3.26a. The transfer function of a Schmitt trigger is shown on fig. 3.26b. The circuit exhibits hysteresis, or backlash. As a result, the output always switches with a fast edge. Figure 3.26d shows a slow changing input and the resulting output.

A Schmitt trigger has the conventional logic symbol with a hysteresis loop similar to the inverter of fig. 3.26c. Schmitt triggers are usually available in inverter, buffer or two-input gate form. The TTL 74123, for example, contains four two-input NAND gates with Schmitt trigger inputs.

Comparison of figs 3.26a and b shows that a Schmitt trigger has better noise immunity than a conventional gate. They are therefore very useful devices for accepting slow changing and possibly noisy signals from the outside world.

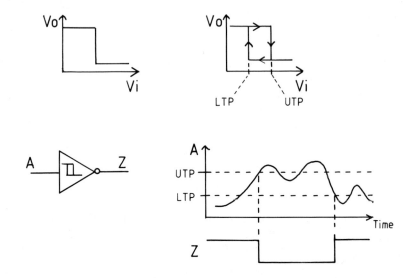

Fig. 3.26 Schmitt trigger. (a) Conventional inverter. (b) Inverter with hysteresis. (c) Symbol. (d) Operation of non-inverting Schmitt buffer.

3.2.12 Open collector and tristate outputs

The TTL NAND gate in fig. 3.27a has a single output transistor rather than the usual totem pole output. The output is connected to Vcc with an external pull-up resistor. Analysis of the gate will show that it behaves like a normal NAND gate (except, of course, that the output rise time is degraded by stray capacitance). This arrangement is known as an open collector output.

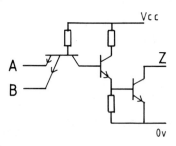 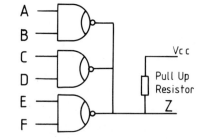

Fig. 3.27 Open collector gates. (a) Open collector circuit. (b) Wire-AND function.

In fig. 3.27b the outputs of several open collector gates are paralleled to a single pull-up resistor (a half-circle on a gate output is sometimes, but by no means universally, used to denote an open collector output). The output Z will be high only if *all* the paralleled gates have high outputs. Using the positive logic convention, the paralleled output behaves as a positive AND gate.

The circuit can also be described by the output Z being low if any gate output is low. The paralleled output therefore also behaves as a negative OR gate. (We should expect this from the description of dualism in section 3.2.8).

Depending on the application, fig. 3.27b can be called a Wire-AND gate or a Wire-OR gate. It has the advantage of giving a possibly complex logic function for the expense of one resistor. Its disadvantage is a poor rising edge, slow speed and a slight degradation of noise immunity.

Open collector outputs are also a useful way of changing logic levels. The external pull-up resistor can be returned to a voltage other than Vcc. If the resistor is returned to, say, 12 volts, the 0/3.5 volt logic levels of TTL can be changed to 0/12 volts for CMOS.

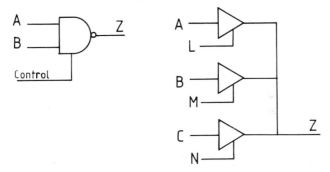

Fig. 3.28   Tristate gates. (a) Tristate NAND. (b) Tristate data selection.

The term tristate gate is somewhat misleading, as a tristate gate does not have three logic levels. It has the two normal logic levels for its family, and a third state where its output goes to a high impedance state. A tristate gate has normal inputs and a separate control input which enables the gate or puts the output to the third high impedance state. Figure 3.28a shows a two-input tristate NAND gate. With the control input high, output Z follows inputs AB as a normal NAND gate. With the control input low, the gate output goes to a high impedance state.

Tristate gates are used to route information from several sources. On fig. 3.28b the outputs of three tristate buffers are connected together. With L high (and MN low) data from A is routed to Z. Similarly M routes B to Z and N routes C to Z. Obviously, the control logic must ensure that no two gates are enabled at the same time (i.e. only one of L, M, N high at once). It should be noted that fig. 3.28b is fundamentally different from the Wire-AND circuit of fig. 3.27b.

TTL has a wide range of devices with tristate outputs, inverters, buffers, gates, storage elements. CMOS devices with tristate outputs are (at the time of writing) restricted to the 4502 hex inverter and 4503 hex buffer plus a few other specialist microprocessor-related devices.

Tristate gates and open collector gates are widely used in the design of highway-based systems, a topic discussed in more detail in chapter 11.

3.3 Combinational design and minimisation

3.3.1 Introduction

The logic designer is usually presented with design requirements in one of three forms: verbal description, Boolean expression, or a truth table. Verbal expressions are often incomplete and vague, but generally take a form such as 'Design a majority vote circuit for five inputs which should all be the same'. It is possible to design simple circuits from such expressions, but it is more often better to convert the verbal description to a Boolean expression or a truth table.

Boolean expressions were described in section 1.4.3, and take a form similar to:

$$Z = A.B.\overline{C} + E.(\overline{A.B} + D)$$

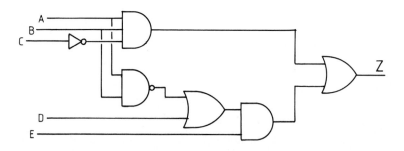

Fig. 3.29 Direct realisation of $Z = A.B.\overline{C} + E.(\overline{A.B} + D)$.

A Boolean expression can be translated directly to a logic circuit; the above expression, for example, is shown in fig. 3.29. A direct translation of a Boolean expression to a circuit does not necessarily lead to the simplest and cheapest circuit, although it will always give a circuit that works.

Truth tables can easily be converted directly to a circuit. Consider this table:

C	B	A	Z	
0	0	0	0	
0	0	1	1	$A \bar{B} \bar{C}$
0	1	0	0	
0	1	1	1	$A B \bar{C}$
1	0	0	0	
1	0	1	0	
1	1	0	0	
1	1	1	1	$A B C$

Z is required to be 1 for three input combinations, namely $A\bar{B}\bar{C}$, $AB\bar{C}$, ABC. The truth table can therefore be directly translated to S of P form, the above example being shown in fig. 3.30a, which is converted to NAND gate form in fig. 3.30b.

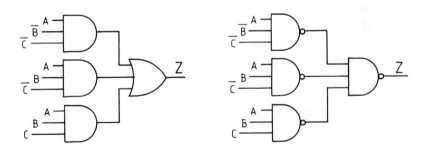

Fig. 3.30 Logic circuits from a truth table. (a) S of P circuit. (b) Equivalent NAND circuit.

In general, any truth table can be converted directly to S of P form by simply noting which input combinations are required to produce a 1 output, and assembling the corresponding S of P circuit. Again, like direct translation of a Boolean expression, this will produce a circuit which will work, but is not necessarily the simplest or most economic solution.

A logic designer should aim to produce a simple and economic circuit. Simple circuits have fewer components, fewer interconnections and are hence more reliable. Classically, logic design was concerned with the minimisation of logic gates. The arrival of ICs has changed this emphasis. The approach now is to design simple and economic circuits with the minimum and cheapest IC packages. MSI and LSI allow complex ICs to be made, so the designer is better using one of eight outputs from a complex IC package (leaving seven-eighths of the package unused) than building the same circuit using four or five simple gate packages.

Minimisation is therefore not as important as it once was. The first step in logic design nowadays should be a search through the range of MSI and LSI chips to see if a device is available which can do the job, be adapted to do the job, or even part of the job. The description of the techniques for minimisation described in the following section should only be used when such a search has failed.

3.3.2 Minimisation using Boolean algebra

All minimisation techniques aim to produce a circuit which uses the minimum number of gates. This can be done by producing an expression in S of P form or P of S form. In general, NAND gate designs are usually done in S of P form and NOR gate designs in P of S form.

The rules of Boolean algebra were described in section 1.4.3. These can be applied to reduce an expression to its minimal form. For example, consider the expression first given in section 1.4.3:

$$Z = AB\bar{D} + A\bar{B}C + \bar{A}B\bar{D} + AD + \bar{B}\bar{C}\bar{D}$$

Grouping the terms with A gives:

$$Z = A(D + \bar{B}C + B\bar{D}) + \bar{A}B\bar{D} + \bar{B}\bar{C}\bar{D}$$

We now note that:

$$D + B\bar{D} = D + B, \text{ giving}$$
$$Z = A(D + B + \bar{B}C) + \bar{A}B\bar{D} + \bar{B}\bar{C}\bar{D}$$

We now note that:

$$B + \bar{B}C = B + C, \text{ giving}$$
$$Z = A(B + C + D) + \bar{A}B\bar{D} + \bar{B}\bar{C}\bar{D}$$

We now observe that $\overline{B}\overline{C}\overline{D}$ is the complement of $B + C + D$. Let Y $= \overline{B}\overline{C}\overline{D}$: then

$$Z = A\overline{Y} + Y + \overline{A}\overline{B}\overline{D}$$

As before $A\overline{Y} + Y = A + Y$, so

$$Z = A + \overline{A}\overline{B}\overline{D} + \overline{B}\overline{C}\overline{D}$$

As before $A + \overline{A}\overline{B}\overline{D} = A + \overline{B}\overline{D}$, giving

$$Z = A + \overline{B}\overline{D} + \overline{B}\overline{C}\overline{D}$$

Again $\overline{B}\overline{D} + \overline{B}\overline{C}\overline{D} = \overline{B}\overline{D}$, so

$$Z = A = \overline{B}\overline{D}$$

Breathing a sigh of relief, we have reached a minimal expression. Note that the C term in this expression is redundant.

Another example:

$$Z = ABC + A\overline{B}.(\overline{\overline{A}.\overline{C}})$$

Applying De Morgan's theorem to the right-hand term gives:

$$Z + ABC + A\overline{B}.(\overline{\overline{A}} + \overline{\overline{C}})$$

but $\overline{\overline{A}} = A$, $\overline{\overline{C}} = C$ giving

$$Z = ABC + A\overline{B}(A + C)$$
$$Z = ABC + AA\overline{B} + A\overline{B}C$$
$$Z = ABC + A\overline{B} + A\overline{B}C$$

We observe $ABC + A\overline{B}C = AC(B + \overline{B}) = AC$; hence

$$Z = AC + A\overline{B}$$

This is again a minimal form.

A final example uses the fact that a term in an S of P expression can be repeated.

$$Z = ABC + AB\overline{C} + A\overline{B}C + \overline{A}BC$$

Repeating the first term does not affect the expression, since:

$$X + X + X = X$$

We now have:

$$Z = ABC + AB\overline{C} + ABC + A\overline{B}C + ABC + \overline{A}BC$$

Grouping and factoring gives:

$$Z = AB(C + \overline{C}) + AC(B + \overline{B}) + BC(A + \overline{A})$$

But $C + \overline{C} = 1$, $B + \overline{B} = 1$, $A + \overline{A} = 1$, so

$$Z = AB + AC + BC$$

This is the minimal form. This expression is actually a majority vote of ABC.

Unfortunately, minimisation using Boolean algebra is rarely obvious and relies on intuition and experience rather than straightforward rules. The use of multiple levels of inversion makes the process tedious and error-prone. Boolean minimisation is really better suited to examination questions than practical design. It is usually far easier to transfer the original Boolean expression to a Karnaugh map and proceed as the following section.

3.3.3 Minimisation using Karnaugh maps

Karnaugh maps were introduced in section 1.4.4. They are a two-dimensional representation of a truth table, in which the map is drawn such that only one variable changes between adjacent cells both vertically and horizontally.

On fig. 3.31, for example, square J represents ABCD.
Moving up to L gives $ABC\overline{D}$, i.e. D only has changed.
Moving down to N gives $AB\overline{C}D$, i.e. C only has changed.
Moving to the right to K gives $\overline{A}BCD$, i.e. A only has changed.
Moving to the left to M gives $A\overline{B}CD$, i.e. B only has changed.
Similar results would be obtained from any cell.

Each cell on the map represents one line on a truth table, and is called a minterm. The required output, 0 or 1, is entered on the map from a truth table or Boolean expression.

For example, the last expression we laboriously minimised using Boolean algebra was:

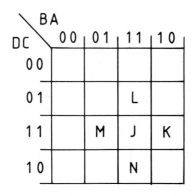

Fig. 3.31 Relationship of adjacent squares on a Karnaugh map.

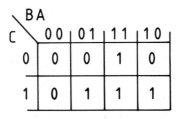

Fig. 3.32 Plot of Z = A.B.C. + A.B.$\bar{C}$ + A$\bar{B}$C + $\bar{A}$BC.

$$Z = ABC + AB\bar{C} + A\bar{B}C + \bar{A}BC$$

This is plotted on the map of fig. 3.32.

In a three-variable map such as fig. 3.32, each cell represents some combination of the three variables. On a four-variable map such as fig. 3.31 each cell represents some combination of the four variables.

Groups of two adjacent cells on a three-variable map represent some combinations of *two* of the thee variables. On fig. 3.33a, groupings for AB and $\bar{B}$C are shown. This map represents:

$$Z = AB + \bar{B}C$$

Two adjacent cells on a four-variable map represent some combination of three of the four variables. On fig. 3.33b, groupings for $\bar{A}$BC, B$\bar{C}\bar{D}$, A$\bar{B}\bar{D}$ and $\bar{B}$CD are shown. This map represents:

$$Z = \bar{A}BC + B\bar{C}\bar{D} + A\bar{B}\bar{D} + \bar{B}CD$$

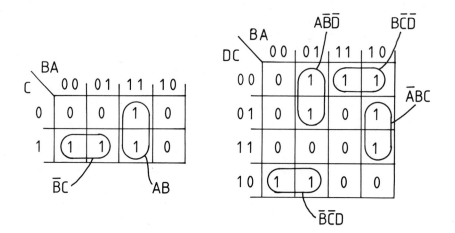

Fig. 3.33 Grouping of two adjacent cells. (a) Three-variable. (b) Four-variable.

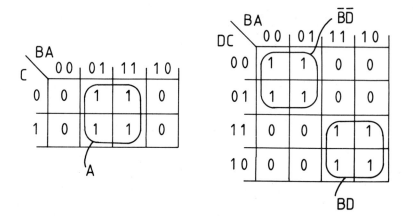

Fig. 3.34 Grouping of four adjacent cells. (a) Three-variable. (b) Four-variable.

Groups of four adjacent cells on a three-variable map represent a single variable. The group on fig. 3.34a represents the variable A, hence

$$Z = A$$

Groups of four adjacent cells on a four-variable map represent some combination of two of the four variables. The groups on fig. 3.34b represent $\overline{B}\overline{D}$ and BD. The map represents:

$Z = \overline{B}\,\overline{D} + BD$

A group of eight adjacent cells on a four-variable map represent a single variable. The groups on fig. 3.35 represent C and $\overline{B}$, so:

$Z = C + \overline{B}$

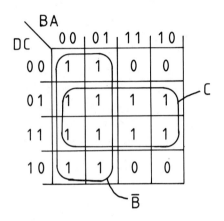

Fig. 3.35 Grouping of eight adjacent cells.

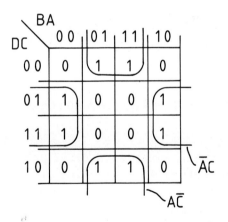

Fig. 3.36 Adjacency of top, bottom and sides.

It is important to realise that top and bottom edges are considered adjacent, as are right and left sides. Groupings can therefore be made around the tops and sides as in fig. 3.36 which represents:

$Z = \overline{A}C + A\overline{C}$

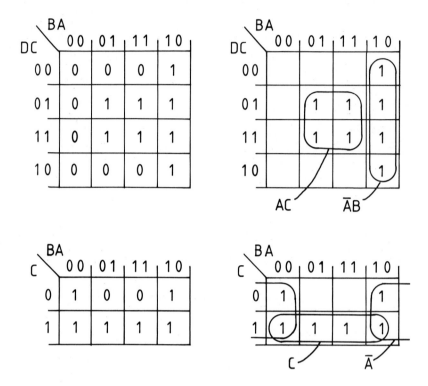

Fig. 3.37 Grouping for minimal expression. (a) Four-variable. (b) Minimal grouping. (c) Three-variable. (d) Minimal grouping.

Consider the expression:

$$Z = \bar{A}\bar{B}\bar{C}\bar{D} + \bar{A}BC\bar{D} + ABC\bar{D} + A\bar{B}C\bar{D} + A\bar{B}CD + ABCD + \bar{A}BCD + \bar{A}B\bar{C}D$$

This has eight terms which are plotted on fig. 3.37a. Using the grouping ideas outlined above we can group the cells as shown in fig. 3.37b. The two groups are AC and $\bar{A}$B. The rather horrendous expression for Z simply becomes:

$$Z = AC + \bar{A}B$$

The same result could, of course, have been found by Boolean algebra.

Figure 3.37c represents:

$$Z = AC + \overline{A}B + \overline{A}\,\overline{B}\,\overline{C} + \overline{A}\overline{B}C$$

By drawing the grouping as in fig. 3.37d, we obtain the two groups C and $\overline{A}$. Z is therefore more simply represented by:

$$Z = \overline{A} + C$$

Note on fig. 3.37d that groups can overlap and go round edges.

A truth table can also be converted to a minimal logic. For example, suppose a truth table gives Z = 1 for the following conditions:

D	C	B	A
0	0	0	0
0	0	0	1
0	0	1	0
0	1	0	0
0	1	0	1
1	0	0	0
1	0	1	0

and Z = 0 for the remaining nine combinations of ABCD. This is plotted on the Karnaugh map of fig. 3.38. Grouped as shown, the minimal expression is:

$$Z = \overline{A}\overline{C} + \overline{B}\overline{D}$$

Note that again the grouping overlaps and goes round the edges.

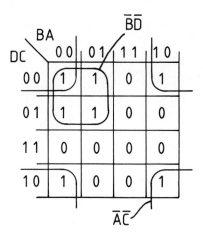

Fig. 3.38 Plotting from a truth table.

Karnaugh maps are a very simple way of producing a minimal logic expression. The rules for doing this are as follows:

(a) Plot the Boolean expression or truth table onto the Karnaugh map.
(b) Form new groups of 1s on the map. Groups must be rectangular and contain 1, 2, 4 or 8 cells. To minimise the expression the groups should be as large as possible and there should be as few groups as possible. Do not forget overlaps and possible round-edge groupings.
(c) From the map, read off the expression for each group formed. The minimal expression is then expressed in S of P form by summing the groups.

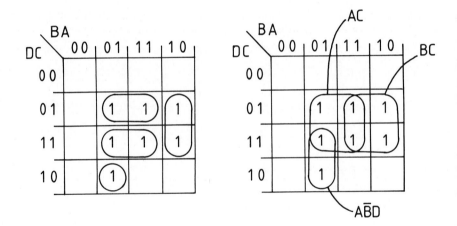

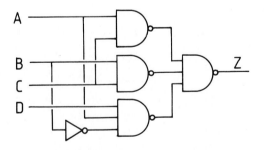

Fig. 3.39 From Boolean expression to minimal circuit. (a) Original plot. (b) Minimal grouping. (c) Minimal circuit.

A typical map is shown in fig. 3.39a with a non-optimal grouping. The expression is:

$$Z = AC\bar{D} + \bar{A}BC + ACD + A\bar{B}\bar{C}D$$

which is logically accurate, but not minimal. The minimal grouping, shown on fig. 3.39b, uses overlaps to form large groups of 4, 4 and 2. The minimal grouping is:

$$Z = AC + BC + A\bar{B}D$$

A Karnaugh map yields a minimum expression in S of P form. It was shown in section 3.2.7 that an S of P expression can be implemented directly with NAND gates. The implementation of the above expression is shown on fig. 3.39c.

3.3.4 The inverse function
The inverse, or complementary, function of an expression generates the inverse of the expression for all input combinations. The Karnaugh map of fig. 3.40b is the inverse of fig. 3.40a (and, of course, fig. 3.40a is the inverse of fig. 3.40b).

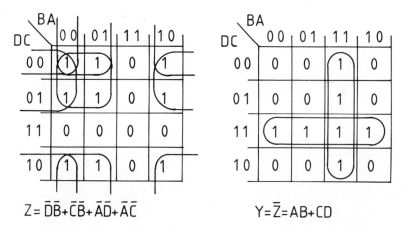

Fig. 3.40 Inverse function. (a) Original function. (b) Inverse function.

Occasionally, the inverse function gives a better minimisation. The Karnaugh map of fig. 3.40a gives the circuit of fig. 3.41a, whilst the inverse map of fig. 3.40b gives the simple circuit of fig. 3.41b. Inverting the inverse function gives the original function again, so adding an inverter to fig. 3.41b gives a simpler version of fig. 3.40a.

If a straightforward minimisation using the Karnaugh map of the original function seems to be unduly complex it is often worth trying

76

the inverse function. If this gives a simple solution, the circuit for the inverse function followed by an inverter will give the original function again.

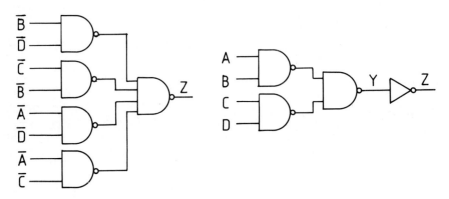

Fig. 3.41 Inverse circuit. (a) Original circuit. (b) Inverse circuit.

In section 3.1 it was mentioned that all possible input combination might not be used. The 'don't care' states can be used to advantage in a Karnaugh map, as we can choose whether they are to be used as 1s or 0s. Suppose was have this truth table:

D	C	B	A	Z
0	0	0	0	0
0	0	0	1	1
0	0	1	0	0
0	0	1	1	0
0	1	0	0	X
0	1	0	1	1
0	1	1	0	1
0	1	1	1	0
1	0	0	0	X
1	0	0	1	X
1	0	1	0	0
1	0	1	1	0
1	1	0	0	X
1	1	0	1	1
1	1	1	0	1
1	1	1	1	0

where X denotes that this input combination cannot occur, allowing us to choose Z to suit the minimisation. The truth table is drawn as in fig. 3.42a. In choosing the state of the 'don't care' states we should aim to form groups as large as possible. Rearrangement as in fig. 3.42b gives two groups of four and the minimal circuit of fig. 3.42c.

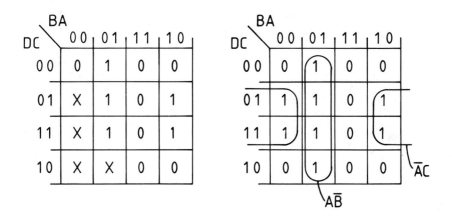

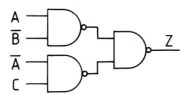

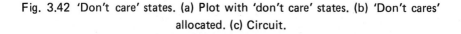

Fig. 3.42 'Don't care' states. (a) Plot with 'don't care' states. (b) 'Don't cares' allocated. (c) Circuit.

Note that assigning 1s and 0s to 'don't care' states effectively defines what the circuit will do under fault conditions. If these fault conditions could cause problems or, in industrial control, some unsafe occurrence the designer should rearrange the 'don't care' allocations.

3.3.5 NOR gate circuits
Karnaugh maps give an S of P solution which translates directly and simply to a NAND gate circuit. If the designer is using NOR gates, his original expression is probably in P of S form, and a NOR gate solution is required. There are two possible approaches. In both, the first

step is to draw up a truth table of the original expression however provided (i.e. P of S form, or S of P form).

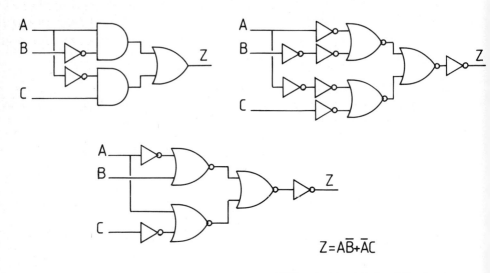

$$Z = A\bar{B} + \bar{A}C$$

Fig. 3.43 Conversion of S of P circuit to NOR circuit. (a) S of P circuit. (b) Direct NOR circuit. (c) Simplified NOR circuit.

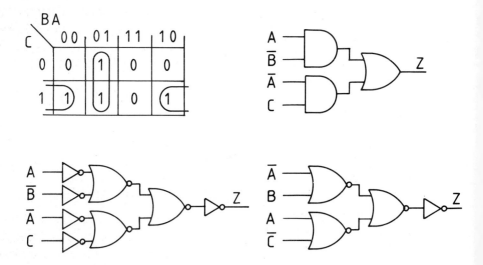

Fig. 3.44 From Karnaugh map to NOR gate circuit. (a) Karnaugh map. (b) S of P expression. (c) NOR gate equivalent. (d) Simplified NOR circuit.

In the first approach a normal Karnaugh map is drawn and minimised to give an S of P solution. This is drawn in AND/OR form, then converted to NOR gates using the circuits of fig. 3.14. For example, we have an S of P circuit on fig. 3.43a. This is converted to NOR gate form on fig. 3.43b which (dropping double inversions) gives fig. 3.43c.

The second approach plots the complementary function, which gives a complementary S of P version of the original expression. This is then converted to NOR gates as above. We now have a complementary version of the function with an inverter on the output. Dropping the inverter gives the required function. (It can be shown, by Boolean algebra, that this technique generates a P of S expression.)

For example, we have a Karnaugh map in fig. 3.44a. Using the first approach we form the S of P circuit as in fig. 3.44b which gives the NOR gate circuit of fig. 3.44c, which in turn simplifies to fig. 3.44d.

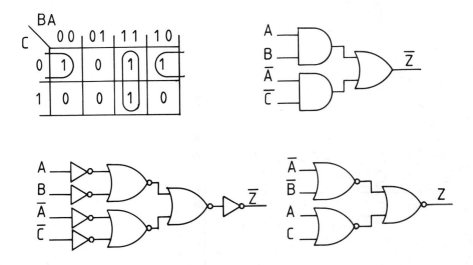

Fig. 3.45 NOR gate circuit using inverse function. (a) Karnaugh map of inverse function. (b) S of P circuit for $\overline{Z}$. (c) NOR circuit for $\overline{Z}$. (d) Simplified NOR circuit for Z.

Using the second approach we form the complementary of fig. 3.45a, which gives the S of P circuit for the complementary function of fig. 3.45b. This is converted to NOR gate form in fig. 3.45c. Dropping the final inverter gives the original function as fig. 3.45d. It will be left as an exercise for the reader to check (either by truth table or Boolean algebra) that figs 3.44d and 3.45d are equivalent.

3.3.6 Exclusive OR gates and Karnaugh maps

If a Karnaugh map appears, in part or whole, as a chequer-board, the use of exclusive OR gates should be considered. Figure 3.46a shows a typical chequer-board pattern and fig. 3.46b the equivalent XOR circuit. A typical Karnaugh map requiring XOR gates and other gates is shown in fig. 3.47a with the equivalent circuit in fig. 3.47b.

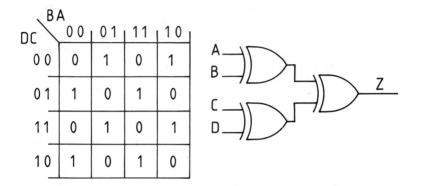

Fig. 3.46 Plot of XOR gate trees. (a) Karnaugh map. (b) Circuit.

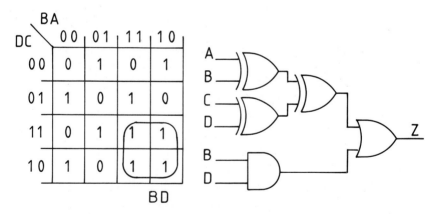

Fig. 3.47 Mixed XOR/AND/OR logic. (a) Required function. (b) Circuit.

3.3.7 Minimisation of five and more variables

Karnaugh maps drawn so far have used three or four variables. Most combinational logic design uses four or fewer variables, or can be split into blocks, each of which uses four or fewer. It is possible, though, to use Karnaugh maps with up to six variables.

With five variables, two maps are drawn; one for the fifth variable and one for its complement. Each map is minimised separately to give its own S of P expression. The two minimal expressions are combined, and examined for common terms. Boolean algebra can then be used to simplify the combined expression by reducing items with common terms. The technique is illustrated in fig. 3.48.

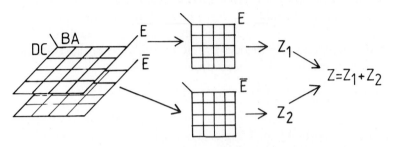

Fig. 3.48 Five-variable Karnaugh map.

Six variables use four maps. The technique is identical to the five-variable approach. Each map is minimised and converted to its S of P expression. The four S of P expressions are combined, and then examined for common terms. Boolean algebra is then used to reduce the common terms. The technique is illustrated in fig. 3.49.

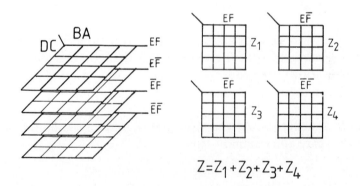

Fig. 3.49 Six-variable Karnaugh map.

If more than six variables are needed, an attempt should be made to break the problem down into manageable blocks. A seven-variable problem, for example, is the sum of two six-variable problems; one gated with the seventh variable and one gated with the seventh variable's complement:

$$Z(ABCDEFG) = G. f(ABCDEF) + \bar{G}. f(ABCDEF)$$

Alternatively, Boolean algebra can be used to produce an expression, possibly non-optimal but logically correct, which can be directly translated into a circuit.

If the designer has access to the necessary programming equipment, or is designing for large production runs, the use of ROMs, UCLAs or PALs should be considered where many variables are encountered. These are described in sections 3.6 and 3.7. Unfortunately these versatile devices are at present not really suitable for use by the amateur as they need programming.

3.3.8 Conclusion

The techniques outlined above yield minimal, or near-minimal, circuits for combinational logic. The designer should, however, be aware of the possible limitations:

(a) MSI and LSI ICs often yield a simpler and more cost-effective solution, so always check the IC catalogue for a ready-made chip first.
(b) Minimisation always gives an S of P or P of S solution. Sometimes this gives gates with many inputs (e.g. a six-input NAND gate is required). Designers are often limited in the range of available gates, and may not be able to build the minimal circuit directly.
(c) Minimisation may disguise a circuit's function, and make fault finding more difficult for people other than the designer.
(d) Minimisation can lead to problems with 'glitches', described further in section 3.4.

The logic designer should therefore use the minimisation techniques above with care and common sense.

3.4 Hazards, races and glitches

Gate propagation delays were introduced in section 1.5.2. In some circumstances these delays can cause unwanted pulses to appear in combinational logic circuits. These unwanted pulses are known variously as hazards, races or glitches.

The logical output of the circuit shown in fig. 3.50a is zero since:

$$Z = A.\overline{A} = 0$$

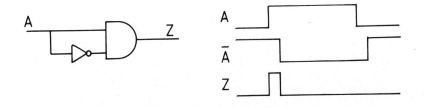

Fig. 3.50 AND gate glitch. (a) Circuit. (b) Timing waveforms.

In practice, however, $\overline{A}$ will be delayed by the propagation delay of the inverter, giving the possible waveforms of fig. 3.50b. As A changes from 0 to 1 a small pulse appears at the output.

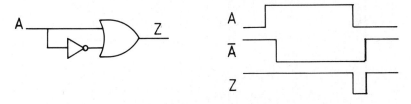

Fig. 3.51 OR gate glitch. (a) Circuit. (b) Timing waveforms.

A similar problem can occur with the OR gate circuit of fig. 3.51a. As A changes from 1 to 0 a small pulse appears at the output.

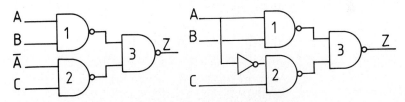

Fig. 3.52 Combinational glitch.

Glitches are not always obvious. The relationship

$$Z = AB + \overline{A}C$$

can obviously be implemented by the circuit of fig. 3.52a. More completely, $\overline{A}$ must be obtained via some inverter as fig. 3.52b. The circuit is logically correct, but if B = C = 1 the output gate 3 is effectively gating A and $\overline{A}$ in the same way as in figs 3.50 and 3.51. If B and C are both 1 and A changes, there is a real possibility that a glitch may occur at the output.

Plotting fig. 3.52 on a Karnaugh map gives fig. 3.53, and an indication of a way to identify, and cure, glitches. There are two groups on the map: AB and $\overline{A}C$. Moving between AB = 11 and AB = 01 we move between groups. This corresponds to A changing from 1 to 0 or 0 to 1. A potential glitch on a Karnaugh map has adjacent 1s not covered by the same group.

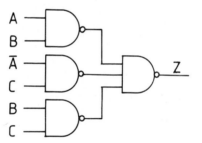

Fig. 3.53 Glitch-free design. (a) Original grouping. (b) Glitch-free grouping.

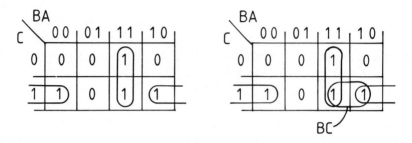

Fig. 3.54 Glitch-free circuit.

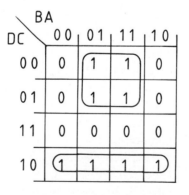

Fig. 3.55 Non-obvious glitch.

To remove the glitch we need to add an additional group as in fig. 3.53b. There are now no adjacent 1s not in the same group. The equivalent circuit is shown on fig. 3.54. Note that the term BC is logically redundant, but prevents glitches as A changes state with $B = C = 1$. Glitch-free circuits are often non-minimal.

In identifying possible glitches the reader should remember that top and bottom edges and left and right sides are adjacent. Figure 3.55 shows a non-obvious potential glitch over the top and bottom edges.

Glitches may not be important. In general, if the output of a glitch-prone circuit is not feeding directly (or indirectly) a storage device, a counter, a monostable or a timer, the glitches will usually have no effect.

If glitches cannot be tolerated, they can be designed out by the technique above (and the loss of a minimal circuit accepted). Alternatively the output signal can be inhibited during changes by a strobe circuit as in fig. 3.56. This is, in reality, a clocked synchronous system, a topic discussed further in sections 7.4 and 4.5.

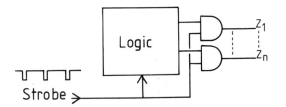

Fig. 3.56 Strobed glitch-free outputs.

Different logic families have different propensities for generating (and ignoring) glitches. The important factor is the relationship between edge speeds and propagation delays. CMOS, with edge speeds similar to, or longer than, the gate propagation delays has a useful tendency to ignore glitches. ECL has very fast edge speeds and tends to be rather glitch-prone.

Glitches can cause intermittent and hard to find faults. The routes by which a glitch is generated in a large system may not be obvious, and the whole system too big to analyse with a Karnaugh map. Large logic systems are usually designed to operate in a clocked manner, a topic discussed further in sections 7.4 and 4.5.

3.5 Data selectors and decoders

Two groups of MSI chips known as data selectors and decoders can be used to implement combinational logic designs. Although intended for other purposes, they can often provide a very economical and simple approach to logic design. Understanding of the techniques below requires an elementary knowledge of binary counting. If the reader is not acquainted with binary, it is suggested that this section be left until chapter 6 has been read.

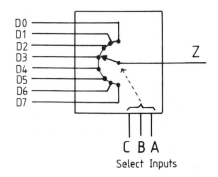

Fig. 3.57 A data selector.

A data selector (also known as a multiplexer) is similar to a rotary switch. On fig. 3.57, there are eight data inputs labelled D0 to D7 connected to an electronic eight-position switch. The switch position is controlled by the inputs CBA which are interpreted as a binary number. If, for example, CBA is 101, D5 will be connected to Z.

When a data selector is used to replace combinational logic, the inputs are connected to ABC and the data inputs of the multiplexer are wired to 1 or 0 to give the required output.

For example, consider the truth table:

C	B	A	Z	Selected D
0	0	0	0	0
0	0	1	1	1
0	1	0	0	2
0	1	1	1	3
1	0	0	1	4
1	0	1	0	5
1	1	0	0	6
1	1	1	1	7

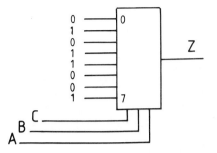

Fig. 3.58 Combinational logic with a data selector.

This can be implemented by the simple one-IC circuit of fig. 3.58. D inputs 1, 3, 4, 7 are wired to a 1 (usually Vcc) and inputs 0, 2, 5, 6 are wired to a 0 (usually 0 V). As the inputs ABC vary, the selected D input gives the required output.

An eight-way data selector can actually be used to replace combinational logic with four inputs, say ABCD. A study of a truth table with four inputs ABCD (and hence 16 entries) will show that for a given combination of three of the inputs (say ABC) the output Z can follow one of four forms:

(i) Always 1

(ii) Always 0

(iii) Follows the fourth input D

(iv) Follows the complement of D (i.e. $\bar{D}$)

This is best illustrated by the 16-line truth table below. Note that this is first drawn up in the usual fashion, then redrawn to group entries with the same value of ABC. This is implemented with one data selector and one inverter as in fig. 3.59.

Any required four-input combinational logic circuit can be built with an eight-way multiplexer by connecting three of the inputs to the select line, and the multiplexer inputs to 1, 0, D or $\bar{D}$ as determined by the truth table.

Multiplexers can be cascaded to cater for more inputs. Suppose we have five inputs ABCDE. This can be represented by:

$$Z = D.E.f_1(A,B,C) + D.\bar{E}.f_2(A,B,C) + \bar{D}.E.f_3(A,B,C) + \bar{D}.\bar{E}.f_4(A,B,C)$$

where f_1 to f_4 (A,B,C) are some logical combinations of ABC. Using the techniques outlined above, f_1 to f_4 can then be further gated by another four-way multiplexer with select lines D, E. The resulting circuit, shown in fig. 3.60, is very quick to design. Usually two four-way multiplexers are contained in one IC, so the circuit uses only 2½ IC packages.

| *Normal* | | | | | *Redrawn* | | | | | | |
D	C	B	A	Z	D	C	B	A	Z	Z is	Selected line
0	0	0	0	0	0	0	0	0	0	D	0
0	0	0	1	0	1	0	0	0	1		0
0	0	1	0	1	0	0	0	1	0	0	1
0	0	1	1	1	1	0	0	1	0		1
0	1	0	0	1	0	0	1	0	1	1	2
0	1	0	1	0	1	0	1	0	1		2
0	1	1	0	1	0	0	1	1	1	1	3
0	1	1	1	1	1	0	1	1	1		3
1	0	0	0	1	0	1	0	0	1	$\bar{D}$	4
1	0	0	1	0	1	1	0	0	0		4
1	0	1	0	1	0	1	0	1	0	D	5
1	0	1	1	1	1	1	0	1	1		5
1	1	0	0	0	0	1	1	0	1	1	6
1	1	0	1	1	1	1	1	0	1		6
1	1	1	0	1	0	1	1	1	1	$\bar{D}$	7
1	1	1	1	0	1	1	1	1	0		7

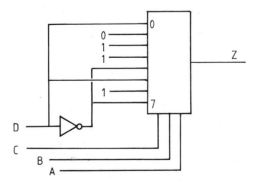

Fig. 3.59 Four-variable combinational logic with a data selector.

TTL has many multiplexers. The 74150 is a 16-way multiplexer (with four select lines) and can be used for five input circuits. The 74151 is an eight-way multiplexer (used for four input circuits).

Tri-state outputs are available on many multiplexers such as the 74LS353 dual four-way multiplexer. Many multiplexers incorporate an enable input which gives further flexibility.

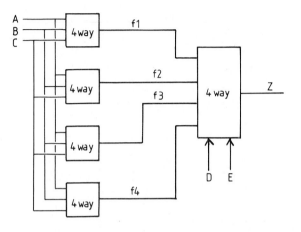

Fig. 3.60 Five-variable combinational logic.

CMOS has both analog multiplexers (known as transmission gates) and digital multiplexers. Transmission gates are described further in section 12.2, but it is sufficient to note that analog multiplexers can be used to switch digital signals. The 4067 is a 16-way analog multiplexer. An eight-way multiplexer is available with the 4513. A dual four-way multiplexer is also available with the 4539.

Summary details of data selectors are given in section 3.8 at the end of this chapter.

A decoder is a device that takes a binary input and converts it to a form of decimal output. A three-bit decoder will have eight output lines, labelled 0 to 7. When ABC are 101, the '5' output line will be at 1, and all other lines are at 0. A four-input decoder will have four inputs and 16 output lines. Again, one selected line will be 1, all others being at 0. Figure 3.61 shows a three-input decoder.

Each output from a decoder corresponds to one line on a truth table. For example, the truth table used for fig. 3.58 has a 1 for binary inputs 1, 3, 4, 7. This can be implemented by the circuit of fig. 3.62a. In practice, most decoders have a *low*-going output (i.e. the selected output is 0, all others are 1). This allows the OR function to be performed by a NAND gate as in fig. 3.62b.

For a single circuit, data selectors are usually more economical. Decoders are very useful, however, where several different outputs are required from a common set of inputs. Figure 3.63 shows several outputs derived from a common set of four inputs.

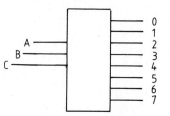

Fig. 3.61 Three to eight line decoder.

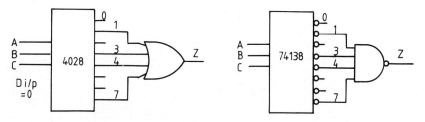

Fig. 3.62 Combinational logic with decoders. (a) Active-high output decoder. (b) Active-low output decoder.

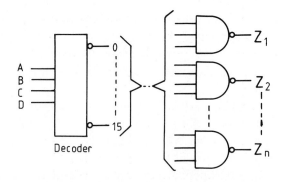

Fig. 3.63 Multiple output combinational logic with decoders.

CMOS provides two four-bit decoders, both with optional storage. The 4514 has high (1) outputs. The 4515 has low (0) outputs. The TTL equivalent is the 74154, unfortunately only available in normal or L form. The output is low-going.

All data selectors and decoders use minimal logic internally. Glitches may occur on the outputs of circuits using them. The importance of these glitches will depend on the application.

3.6 ROM and PROM logic

ROM stands for Read Only Memory, and PROM for Programmable Read Only Memory. These devices are designed for use in micro-computers to hold fixed programs or fixed data, but they can prove a very cheap and versatile way of implementing large logic functions. Unfortunately they are best suited to medium/large productions runs, and are not really applicable to one-off designs.

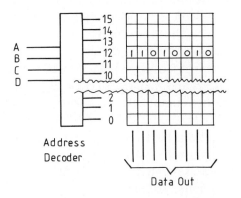

Fig. 3.64 Combinational logic with ROMs.

A ROM or PROM can be represented by fig. 3.64. Internally the device has many locations, each of which holds a preselected binary pattern, loaded, at the design stage, by the user. A binary address selects one location, whose contents appear at the output. Figure 3.64 is a 16 by 8 ROM, meaning it has 16 locations, each of which holds 8 bits. If the input address is 12, for example, the output data would be 11010010.

A ROM can be considered as a hard-wired truth table. The 16 by 8 ROM of fig. 3.64 will give eight independent outputs for four input signals connected to the address line.

Commercially available ROMs and PROMs are generally much larger than the simple example above. Typical is the 2708 PROM organised as 1024 by 8 (i.e. 1024 locations, each location holding eight bits). This has nine input lines, so it can provide eight different output combinations of the nine inputs.

A ROM, or more correctly a Mask Programmed ROM, is provided ready-loaded to the designer's truth table by the manufacturer. The ROM is best suited to large systems with large production runs.

A PROM can be loaded by the user with a device called a PROM programmer. Some PROMs (called EPROMS) can be erased by ultraviolet light and re-programmed. EPROMs are very useful at the design stage where the inevitable oversights can be quickly and cheaply corrected.

3.7 UCLAs, PALs and PLAs

These devices are close relatives of ROMs, but are specifically designed for use in logic systems. UCLA stands for Uncommitted Logic Array (also known as ULAs and gate arrays). An integrated circuit consists of a small slice of silicon in which is etched the various components making the gates, etc. These are interconnected by a thin metallised layer to form the required logic function.

A UCLA consists of an IC which contains a large number of assorted gates, buffers and memories (typically 1000-10000 devices). The user specifies his circuit, which is formed by the design of the metallised layer. The IC slice is standard to all users, the metallisation to one user's application. Because the slice is expensive to design, and the metallisation relatively cheap, UCLAs allow designers to have their 'own' IC at a reasonable price. UCLAs are, however, only cost-effective for large volume production.

PALs (Programmable Array Logic, also known as PLAs by juggling the letters) are rather like do-it-yourself UCLAs. Internally they contain a matrix of true/complement inputs and output AND/OR combinations arranged as in fig. 3.65. Each crossing point is initially linked, and is 'blown' open by the designer (using a programming device) to leave his desired function. Inherently a PAL gives an S of P expression directly. PALs are available with tristate outputs and latched tristate output for use in clocked systems. Figure 3.65 is contained in a 20-pin package.

3.8 Gates and other combinational logic ICs

3.8.1 TTL

Inverters		Availability
7404	Hex inverter	ALL
7405	Hex inverter open collector	N, H, LS, S

AND gates		
7408	Quad two-input AND	N, H, LS, S
7409	Quad two-input AND open collector	N, LS, S
7411	Triple three-input AND	H, LS, S

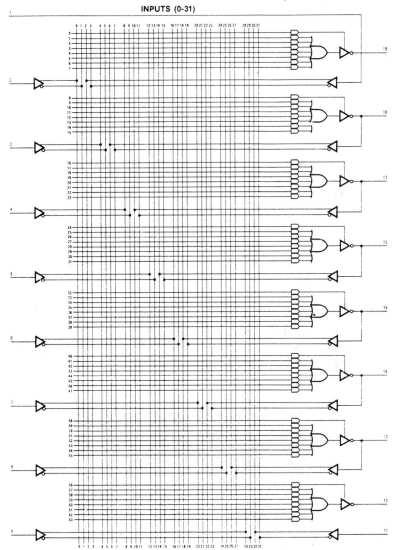

Fig. 3.65 Typical PAL.

AND gates (continued)

7415	Triple three-input AND open collector	H, LS, S
7421	Dual four-input AND	H, LS

OR gates

7432	Quad two-input OR	N, LS, S

NAND gates

7400	Quad two-input NAND	ALL
7401	Quad two-input NAND open collector	N, H, LS
7403	Quad two-input NAND open collector	N, L, LS, S
7410	Triple three-input NAND	ALL
7412	Triple three-input NAND open collector	N, LS
7420	Dual four-input NAND	ALL
7422	Dual four-input NAND open collector	N, H, LS, S
7430	Eight-input NAND	ALL
74133	13-input NAND	S
74134	12-input NAND (tristate)	S

NOR gates

7402	Quad two-input NOR	N, L, LS, S
7423	Expandable Dual four-input NOR	N
7425	Dual four-input NOR with strobe	N
7427	Triple three-input NOR	N, LS
74260	Dual five-input NOR	N

XOR gates

7486	Quad two-input XOR	N, L, LS, S
74136	Quad two-input XOR open collector	N, LS

XNOR gates

74135	Quad Exclusive OR/NOR	S
74266	Quad XNOR open collector	LS

AND-OR (AO)/AND-OR-INVERT (AOI) gates

7450	Expandable two-wide, two-input AOI	N, H
7451	Two-wide, two-input AOI	ALL
7452	Four-wide, two-input AO	H
7453	Expandable four-wide, two- or three-input AOI	N, H
7454	Four-wide, two- or three-input AOI	N, H, L, LS
7460	Expander for 23, 50, 53, 55 dual	

AND-OR (AO)/AND-OR-INVERT (AOI) gates

	four-input	N, H
7461	Expander for 52 triple three-input	H
7462	Expander for 50, 53, 55 3-2-2-3-input	H
7464	Four-wide, 4, 2, 3, 2 input AOI	S
7465	Four-wide, 4, 2, 3, 2 input AOI open collector	S

Schmitt triggers

7413	Dual four-input NAND	N, LS
7414	Hex inverter	N, LS
74132	Quad two-input NAND	N, LS, S

Buffers/inverters (High current, open collector, tristate)

			Volts	*mA*	*Type*
7406	Hex inverting buffer	o/c	30	40	N
7407	Hex buffer	o/c	30	40	N
7416	Hex inverting buffer	o/c	15	40	N
7417	Hex buffer	o/c	15	40	N
7428	Quad two-input NOR		TTL	48	N
			TTL	24	LS
7433	Quad two-input NOR	o/c	5	48	N
		o/c	5	24	LS
7437	Quad two-input NAND		TTL	48	N
			TTL	24	LS
			TTL	60	S
7438	Quad two-input NAND	o/c	5	48	N
		o/c	5	24	LS
		o/c	5	60	S
		o/c	15	48	N suffix A
7440	Dual four-input NAND		TTL	48	N
			TTL	60	H
			TTL	24	LS
			TTL	60	S
74125	Quad buffer		Tristate	–	N, LS
74126	Quad buffer		Tristate	–	N, LS
74128	Quad two-input NOR		TTL	48	N
74240	Octal inverting buffer		Tristate	–	LS, S
74241	Octal buffer		Tristate	–	LS, S
74244	Octal buffer		Tristate	–	LS
74365	Hex buffer		Tristate	–	N, LS
74366	Hex inverting buffer		Tristate	–	N, LS

Buffers/inverters (High current, open collector, tristate)

74367	Hex buffer	Tristate	–	N, LS
74368	Hex inverting buffer	Tristate	–	N, LS
74425	Quad buffer	Tristate	–	N
74426	Quad buffer	Tristate	–	N

Data selectors/multiplexers and decoders

		Availability
7442	BCD to decimal decoder	N, L, LS
74150	16-input data selector	N
74151	Eight-input data selector	N, LS, S
74153	Dual four-input data selector	N, L, LS, S
74154	Four-line to 16-line decoder	N, LS
74157	Quad two-input data selector	N, L, LS, S
74158	Quad two-input data selector inverting	LS, S
74251	Eight-input data selector tristate	N, LS, S
74253	Dual four-input data selector tristate	LS, S
74257	Quad two-input data selector tristate	LS, S
74258	Quad two-input data selector inverting tristate	LS, S
74351	Dual eight-input data selector	N
74352	Dual four-input data selector inverting	LS
74353	Dual four-input data selector inverting tristate	LS

Other decoders may be found in sections 8.14 and 9.8.

Data selectors/multiplexers with latch storage

7498	Quad two-input data selector	L
74298	Quad two-input data selector	N, LS
74354	Eight-input data selector	LS
74355	Eight-input data selector, open collector only	LS
74356	Eight-input data selector with D-type	LS
74357	Eight-input data selector with D-type, open collector only	LS
74398	Quad two-input data selector	LS
74399	Quad two-input data selector	LS

3.8.2 CMOS

Inverters/buffers
4041	Quad true/complement buffer
4049	Hex inverting buffer
4050	Hex non-inverting buffer
4069UB	Hex inverter (unbuffered)
4502	Strobed hex inverter/buffer
40097	Hex non-inverting buffer tristate
40098	Hex inverting buffer tristate
40244	Octal buffer tristate

AND gates
4073	Triple three-input AND
4081	Quad two-input AND
4082	Dual four-input AND

OR gates
4071	Quad two-input OR
4072	Dual four-input OR
4075	Triple three-input OR

NAND gates
4011	Quad two-input NAND
4012	Dual four-input NAND
4023	Triple three-input NAND
4068	Eight-input NAND

NOR gates
4000	Dual three-input NOR plus one inverter
4001	Quad two-input NOR
4002	Dual four-input NOR
4025	Triple three-input NOR
4078	Eight-input NOR

XOR gates/XNOR gates
4030	Quad XOR
4070	Quad XOR
4077	Quad XNOR

AND-OR-INVERT gates (AOI)
4085   Dual two-wide, two-input AOI
4086   Four-wide, two-input AOI

Schmitt triggers
4093   Quad two-input NAND
40106   Hex inverter

Data selectors/multiplexers and decoders
4019   Quad two-input multiplexer
4028   BCD to decimal decoder
4512   Eight-input multiplexer tristate output
4514   1-of-16 decoder with input latches. True output
        (Can also be used as demultiplexer)
4515   1-of-16 decoder with input latch. Complement output
        (Can also be used as demultiplexer)
4519   Quad two-input multiplexer
4539   Dual four-input multiplexer

Analog multiplexers are covered in section 12.6.

CHAPTER 4

Storage

4.1 Introduction

Most logic systems require some form of memory. Even the simple relay motor starter of fig. 4.1 uses a latching relay, RL, to remember that the start button has been pressed. The operation of the motor starter is obvious; when start is pressed RL energises and stays energised via its own contact after start has been released. When stop is pressed, RL de-energises again. The relay is thus a simple storage device remembering which input was last present.

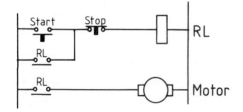

Fig. 4.1 Relay memory.

4.2 Cross-coupled flip-flops

The logical equivalent of fig. 4.1 is the cross-coupled NOR circuit of fig. 4.2. This is shown using half of a CMOS 4001 quad NOR IC package, but could equally well be constructed with any two NOR elements. The two inputs (S, R) stand for Set and Reset. The reason for the output notation (Q, $\overline{Q}$, pronounced Q bar) is lost in the mists of time.

To describe the circuit operation, let us assume both inputs are at 0, and for some reason output Q (pin 3) is at a 1 and output $\overline{Q}$ (pin 4) is at a 0. Gate 'a' thus has pin 2 at a 0, giving the 1 out on pin 3, and gate 'b' has pin 5 at a 1 giving the 0 out on pin 4. The circuit is therefore stable with Q at a 1, and $\overline{Q}$ at a 0.

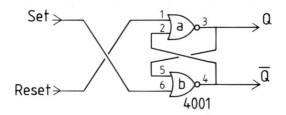

Fig. 4.2 Cross-coupled NOR flip-flop.

If R is now taken to a 1, the output of gate 'a' will go to a 0. Gate 'b' now has both inputs at a 0, so its output will go to a 1. The circuit is now stable with Q at a 0 and $\overline{Q}$ at a 1. The circuit will remain stable even when R is returned to a 0.

If S is now taken to a 1, $\overline{Q}$ will go to a 0, Q back to a 1, and the circuit will remain in this state when S returns to a 0.

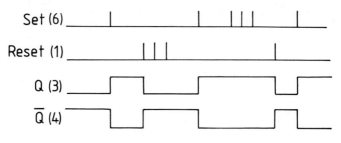

Fig. 4.3 Operation of RS flip-flop.

The circuit is summarised by fig. 4.3, and it can be seen that the circuit remembers which input was last a 1. If Q is at a 1, input S was last at a 1; if $\overline{Q}$ is at a 1, input R was last at a 1. If both inputs are 1 together, both outputs go to a 0. In most applications this is a disallowed state.

Figure 4.3 can, more conveniently, be represented by the simple table:

S	R	Q	$\overline{Q}$	
1	0	1	0	
0	1	0	1	
0	0	outputs hold last state		
1	1	0	0	but normally disallowed

The cross-coupled NOR flip-flop is usually called an RS flip-flop, and can be shown on logic diagrams by the symbol of fig. 4.4.

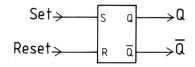

Fig. 4.4 Symbol for NOR RS flip-flop.

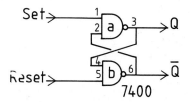

Fig. 4.5 Cross-coupled NAND flip-flop.

A cross-coupled flip-flop can also be constructed with NAND gates as shown for a TTL 7400 on fig. 4.5. The circuit works in a similar manner to fig. 4.2, except that it remembers which input was last at a 0. The quiescent state is thus both inputs at a 1, and both inputs at a 0 is normally disallowed.

Figure 4.5 can be described by the table:

S	R	Q	$\overline{Q}$	
0	1	1	0	
1	0	0	1	
1	1	outputs hold last state		
0	0	1	1	but normally disallowed

The cross-coupled NAND flip-flop is also called an RS flip-flop, and is represented on logic diagrams by the symbol in fig. 4.6. Using the notation outlined in chapter 3 for the invert operation the small circles on the S and R inputs denote that the inputs respond to a 0 state.

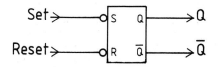

Fig. 4.6 Symbol for NAND RS flip-flop.

Both the NOR and NAND cross-coupled flip-flops are described as RS flip-flops, so there is a possible source of confusion when reading data sheets. For example, there are two quad RS flip-flop ICs available in CMOS; the 4043 and 4044. The 4043 uses cross-coupled NORS, the 4044 uses cross-coupled NANDs.

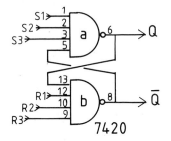

Fig. 4.7 Three-input NAND RS flip-flop.

Multiple input gates can also be used to construct RS flip-flops. Figure 4.7 shows a flip-flop constructed from a 7420 dual four-input NAND. The circuit will go to its set state if any of S1, S2 or S3 are taken to a 0, and will go to its reset state if any of the reset inputs are taken to a 0. As before, a simultaneous set and reset input is normally not allowed.

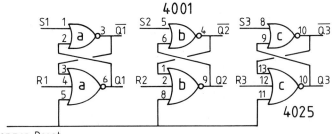

Fig. 4.8 Flip-flops with common reset.

Multiple input gates also allow a common reset to be applied to several flip-flops. In fig. 4.8 we have three NOR RS flip-flops constructed from a 4001 quad two-input NOR and a 4025 triple three-input NOR. Each flip-flop has its own set and reset inputs, but there is also a common reset that puts all three flip flops into a reset state.

A very simple RS flip-flop can be constructed from inverters to memorise the state of contacts on push-buttons and similar items. The CMOS version, using the 4069 hex inverter, is shown in fig. 4.9. PB1 sets $Q = 1$, $\overline{Q} = 0$. PB2 resets to $Q = 0$, $\overline{Q} = 1$. The TTL version is best constructed with open collector inverters such as the 7416 as shown on fig. 4.10. As before, PB1 sets, and PB2 resets the memory.

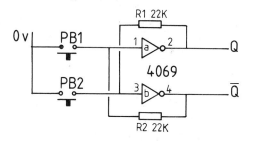

Fig. 4.9 CMOS debouncing flip-flop.

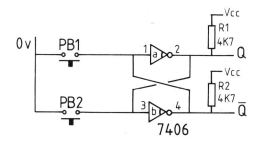

Fig. 4.10 TTL debouncing flip-flop.

RS flip-flops are widely used to remove bounce from external switches, a topic discussed further in section 13.5.2. The cross-coupled inverter is a simple and economical method of constructing a bounce-removing flip-flop.

4.3 Edge triggering

In many schemes it is difficult to avoid the disallowed states on the inputs to an RS flip-flop. In these circumstances it is often more convenient to utilise a simple differentiator on each input as shown in fig. 4.11. These circuits produce a narrow pulse; fig. 4.11a on a negative edge, fig. 4.11b on a positive edge. These pulses can then be used to set and reset a conventional NAND or NOR RS flip-flop. The diodes clip the unwanted opposite sense pulse. Although TTL and CMOS gates do have protection diodes on the input, it is not good practice to rely on them.

Practical CMOS circuits are shown in fig. 4.12. The values shown give an input pulse of about 200 µs, which could be made wider, or narrower, by changing the resistor or capacitor values to suit the repetition rate of the input. Figure 4.12a sets and resets a NOR flip-flop on positive edges, fig. 4.12b sets and resets a NAND flip-flop on negative edges.

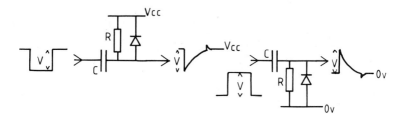

Fig. 4.11 Basic edge-triggering circuits. (a) Negative edge trigger. (b) Positive edge trigger.

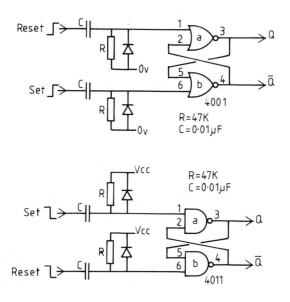

Fig. 4.12 Edge triggered CMOS circuits. (a) Positive edge triggered NOR flip-flop. (b) Negative edge triggered NAND flip-flop.

The equivalent TTL circuits are shown on fig. 4.13. A TTL gate does not have the sharp transfer characteristic of CMOS gates, making it advisable to use Schmitt trigger gates such as the 74132.

There being no Schmitt NOR gate in the TTL range, the NAND version should be used if possible.

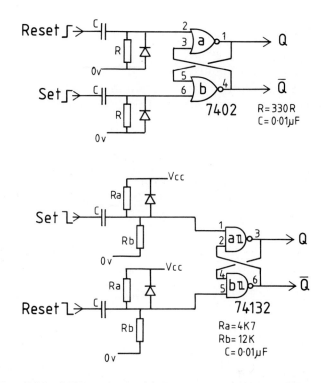

Fig. 4.13 Edge-triggered TTL circuits. (a) Positive edge triggered NOR flip-flop. (b) Negative edge triggered NAND flip-flop.

Edge triggering can provide a solution to many problems, but its use should be tempered with care. Unpredictable results occur if simultaneous set and reset edges occur within the duration of the trigger pulses, so the pulse width should be carefully chosen. A fast, clean edge must be provided in both directions, or mistriggering may occur. It is likely that the non-triggering edges at the driving gates will be degraded by the action of the clamping diodes. Although this is not critical for the RS flip-flop, it may cause problems elsewhere. Finally, RS flip-flops using an RC input are more prone to noise pick-up. It is particularly important to provide adequate supply decoupling and to keep the input leads short.

4.4 Transparent latch

The transparent latch (also known as the follow/hold latch) is a development of the simple RS flip-flop. In its simplest form it has a single data input, shown on fig. 4.14 as A, a single output Q and an enable line. With the enable line at 1, there is effectively no memory, and Q follows the input A. With the enable line at 0, Q holds the state of A prior to the 1-to-0 transition of the enable line. The transparent latch can thus be used to 'freeze' changing data.

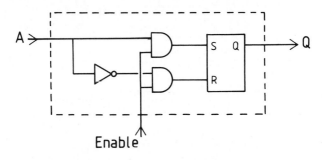

Fig. 4.14 The transparent latch.

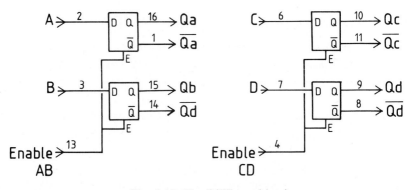

Fig. 4.15 The 7475 quad latch.

Transparent latches are normally found in quad ICs, a typical device being the 7475 shown in fig. 4.15. This has four latches and two enables, each driving two latches. Both Q and $\overline{Q}$ outputs are brought out for each latch. With the enable input at 1, the latch is in its transparent state with Q following D. With the enable input at 0, the last state of D is held. Its action is summarised by:

Data	Enable	Q	$\bar{Q}$
0	1	0	1
1	1	1	0
X	0	Qo	$\bar{Q}$o

The 7475 is shown in a typical application in fig. 4.16, where it is used to freeze the state of a 7490 decade counter for decoding to drive a seven-segment display.

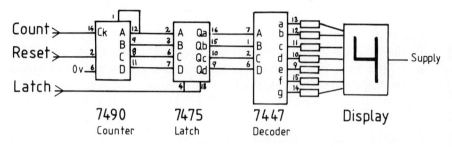

Fig. 4.16 Typical application of transparent latch.

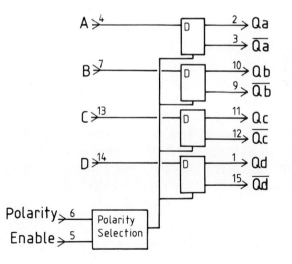

Fig. 4.17 CMOS 4042 transparent latch.

A typical CMOS transparent latch is the 4042 shown in fig. 4.17. This is similar to the 7475 with the addition of the polarity input. This allows the sense of the enable to be changed. Its action is summarised by:

Data	Enable	Polarity	Q	$\bar{Q}$	
1	0	0	1	0	
0	0	0	0	1	
X	1	0	Qo	$\bar{Q}$o	Hold state
1	1	1	1	0	
0	1	1	0	1	
X	0	1	Qo	$\bar{Q}$o	Hold state

Care should be taken when reading data sheets on latches. Some data sheets refer to RS flip-flops as 'latches', and others use the term for the clocked D-type flip-flop described below in section 4.5.3. The use of the term 'clock' for the enable line is also confusing as it could imply the latch is a true clocked device (which it is not).

There are many variations on the transparent latch. Octal latches with three-state output (e.g. 74373) are used for computer highways. The 4511 includes a four-bit transparent latch with a seven-segment decoder.

4.5 Clocked storage

4.5.1 Introduction

If the simple flip-flops of the previous section are used in large schemes, problems are often encountered with timing glitches. On fig. 4.18 we have four-bit data which moves through some logic from storage point A to B to C and so on. Movements similar to this occur in many arithmetic processing schemes. If transparent latches were used, and all the enables were simply connected together, data would shoot through all the points A, B, C simultaneously. To allow data to proceed in an orderly manner, some fairly complex sequencing would be needed to drive each enable individually.

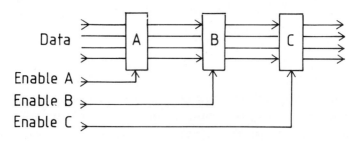

Fig. 4.18 Strobed data transfer.

The concept of clocked memories greatly simplifies the design of schemes similar to fig. 4.18. The simplest clocked memories operate on the master-slave principle shown on fig. 4.19. The master-slave flip-flop consists of two gated flip-flops, strobed by a clock pulse. When the clock input is a 0, the SR inputs are gated into the master flip-flop. At this time the outputs (from the slave flip-flops) are unchanged. When the clock input is 1, the master flip-flop is isolated from the inputs, and the Q, $\bar{Q}$ outputs from the master flip-flop are gated to the slave flip-flop. The device outputs thus change state when the clock input goes from 0 to a 1.

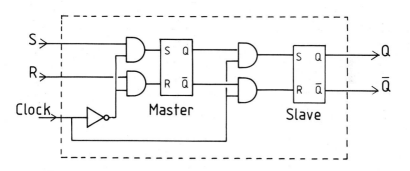

Fig. 4.19 Basic clocked master/slave flip-flop.

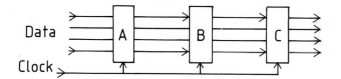

Fig. 4.20 Clocked data transfer.

The orderly transfer of data overcomes the problem outlined in fig. 4.18, and allows a common clock line to drive all the devices, as shown on fig. 4.20.

The device illustrated in fig. 4.19 is a clocked RS flip-flop, and is actually a rather uncommon device. Most clocked flip-flops use either the JK or D-type configuration described below.

4.5.2 The JK flip-flop

The JK flip-flop is simply a clocked RS flip-flop with additional logic to cover the previously disallowed condition where both inputs are 1 when the device is clocked. Under these conditions, the outputs toggle; that is, Q takes the state of $\bar{Q}$ prior to clocking, and $\bar{Q}$ takes the state of Q. This is best described by:

J	K	CK	Q	$\bar{Q}$	
0	0	⎍	no change		Q = previous Q, $\bar{Q}$ = previous $\bar{Q}$
0	1	⎍	0	1	Reset state
1	0	⎍	1	0	Set state
1	1	⎍	Toggle		Q = previous $\bar{Q}$, $\bar{Q}$ = previous Q

The inputs are denoted J and K to avoid confusion with the RS flip-flop.

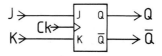

Fig. 4.21 The JK flip-flop.

The logic symbol for a JK flip-flop is shown in fig. 4.21. The above table is often shown:

J	K	CK	Qn+1
0	0	⎍	Qn
0	1	⎍	0
1	0	⎍	1
1	1	⎍	$\overline{Qn}$

Where Qn denotes state prior to clock pulse and Qn+1 denotes state after clock pulse.

4.5.3 The D-type flip-flop

A D-type flip-flop is formed by adding an inverter to the R input of a clocked RS flip-flop as shown in fig. 4.22. The D-type therefore has one data input and a clock input. When the device is clocked, Q takes the state of D:

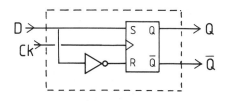

Fig. 4.22 The D-type flip-flop.

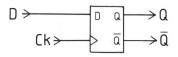

Fig. 4.23 Symbol for D-type flip-flop.

D	CK	Q	$\bar{Q}$
1	⊓_	1	0
0	⊓_	0	1

The logic symbol for a D-type flip-flop is shown in fig. 4.23. Rather confusingly, the term D-type latch is often used.

4.5.4 Level and edge triggered devices

The circuit of fig. 4.19 gates data from the input to the master flip-flop when the clock is at 0 and from the master to the slave flip-flop when the clock is at 1. Because this movement of data is dependent on the clock level, the circuit is called a level-clocked device.

Some care needs to be taken with level-clocked devices to ensure that the data being strobed in is correct, and timing can be problematical in large systems. In particular, changing the input states when the clock is gating data into the master flip-flop may cause unexpected results. Most modern flip-flops include additional logic to transfer the data from input to master and master to slave during clock pulse edges. Usually this is achieved by holding data internally on small capacitors during the clock edge. Devices operating on this principle are more versatile, and are known as edge triggered devices. Level triggering devices should be operated as shown in fig. 4.24.

Most standard TTL JK flip-flops (7473, 7476) are level triggered. All D types (7474), all Schottky, low powered Schottky TTL and all CMOS devices are edge triggered.

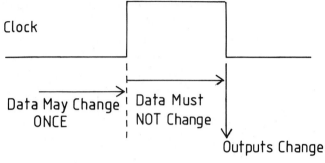

Fig. 4.24 Level-clocked flip-flops.

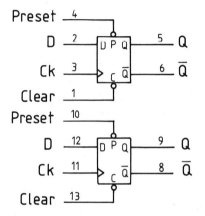

Fig. 4.25 7474 dual D-type flip-flop.

In this book we will use the notation below for clock inputs:

⌐⌐ Negative level triggered; outputs change when clock goes 1 to 0. Input data should not change whilst clock is 1

⌐⌐ Positive level triggered; outputs change when clock goes 0 to 1. Input data should not change whilst clock is 0

↓ Negative edge triggered; outputs change when clock goes 1 to 0

↑ Positive edge triggered; outputs change when clock goes 0 to 1

If possible, edge triggered devices should be used for new designs.

4.5.5 Direct inputs

Most JK and D-type flip-flops have direct inputs in addition to the clocked inputs described in the previous sections. These direct inputs operate in a similar manner to the Set and Reset inputs on the RS flip-flop, and do not require the clock pulse. In most devices the operation of the direct inputs will override the clocked inputs.

A typical TTL flip-flop with clocked and direct inputs is the 7474 dual D-type, shown in fig. 4.25. The direct inputs are known as Preset and Clear, and are operated by 0 logic levels. A 0 to the preset input sets $Q = 1$ and $\bar{Q} = 0$. A 0 to the clear makes $Q = 0$ and $\bar{Q} = 1$. The direct inputs override the clocked inputs. As usual, the small circles on preset and clear denote 0 active inputs.

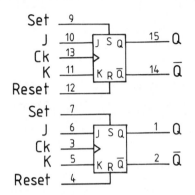

Fig. 4.26 4027 dual JK flip-flop.

The 4027 dual JK flip-flop shown in fig. 4.26 is typical of CMOS devices. The direct inputs, called Set and Reset this time, are operated by 1 logic levels. It will be noted that this is the opposite sense to the direct inputs on the 7474, there being no standard for direct inputs.

4.5.6 Multi-input flip-flops

Some flip-flop ICs utilise multi-inputs for additional flexibility. A typical device is the 74101 shown in fig. 4.27. These devices can simply be considered as a conventional JK flip-flop preceded by the logic gates. The J input is at a 1 if J1A and J1B are both 1 OR J2A and J2B are both one. The K input is similar, so we can write:

$$J = (J1A \times J1B) + (J2A \times J2B)$$
$$K = (K1A \times K1B) + (K2A \times K2B)$$

Multi-input flip-flops are not particularly common devices.

4.5.7 Toggle flip-flops

The toggle, or T-type flip-flop is a special clocked flip-flop used in counters. In its simplest form, shown in fig. 4.28, it has one clock input (denoted by T), two direct inputs, S and R, and the usual two

114

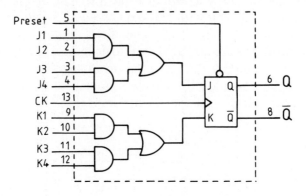

Fig. 4.27 A multi-input flip-flop, 74101 JK.

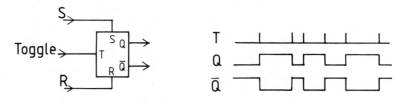

Fig. 4.28 The toggle flip-flop. (a) Logic symbol. (b) Operation.

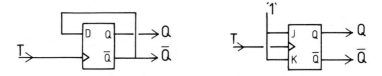

Fig. 4.29 Implementing a toggle flip-flop. (a) Using a D-type flip-flop. (b) Using a JK flip-flop.

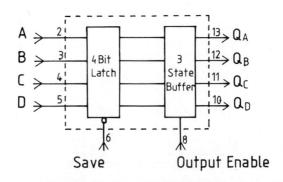

Fig. 4.30 Non-volatile quad latch 9102.

outputs. Each time a clock pulse is applied to the T input, Q and $\bar{Q}$ toggle.

The T-type flip-flop is not available as such in ICs because a D-type or JK flip-flop can be made into a toggle flip-flop. Figure 4.29 shows the D-type and JK flip-flops used in their toggle modes.

The toggle flip-flop is the basis of all counters and is discussed further in section 7.2.

4.6 Non-volatile storage

The TTL and CMOS storage devices discussed above are all volatile devices, i.e. they return to a random state if the power supply is removed and re-applied. In some applications (e.g. holding datum positions in numerically controlled machine tools) this can be very undesirable. With CMOS logic it is feasible to use battery backup, but this is not practical with TTL.

A new generation of storage devices based on Metal Nitride Oxide Silicon (MNOS) technology, uses charge storage to provide non-volatile storage. A typical device is the 9102 quad latch shown in fig. 4.30. The one disadvantage of these devices is the need for two supplies, +5 V and −12 V, although these are standard microprocessor voltages.

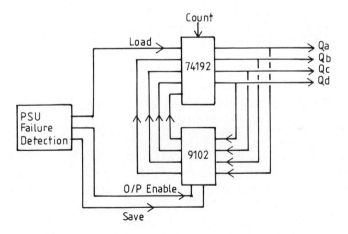

Fig. 4.31 Typical application of non-volatile latch.

A typical application is shown in fig. 4.31. A 74192 up/down counter (see section 7.3.8) is part of a logic scheme. A supply monitoring circuit detects the supply failure, and in the few milliseconds before the 5 volt rail runs down it transfers the 74192 contents to the 9102 latch. On power up, the supply monitoring circuit loads the 9102 contents back to the 74192 via the preset inputs.

4.7 Practical considerations

Flip-flops are designed to store transient events, and as such are sensitive to noise. It is good practice to decouple the supply rails with a 0.01 μF capacitor for *every* flip-flop IC to obviate supply-borne noise.

Inputs are particularly sensitive to noise, and the rules outlined in chapter 13 concerning reflections due to long lines and crosstalk between adjacent tracks should be followed. A less obvious problem can occur via the outputs of simple RS flip-flops. It is possible to set or reset a flip-flop by forcing the outputs. Extreme cases of crosstalk or reflection can force the outputs of a simple RS flip-flop using NANDs or NORs, so the wiring rules should be followed for inputs *and* outputs of simple flip-flops. Most flip-flop ICs, however, use buffered outputs, so noise on output lines is not usually a problem. Flip-flop outputs should not, however, drive long lines directly.

Fig. 4.32 Problems with slow clock edges. (a) Noise on edge. (b) Non-synchronous trigger.

Clock pulses should be clean and fast. A noisy clock pulse as in fig. 4.32a can cause multiple clocking. A slow clock pulse as in fig. 4.32b can result in very peculiar symptoms, as different flip-flops trigger at different points of the clock edge. In general, edge-triggered devices are more tolerant than level-triggered devices.

Synchronous devices have a specified 'set-up and hold time'. These are, respectively, the minimum time that the data must be present before, and after, the active clock edge. This is really only relevant when a device is operated near its limit.

4.8 Storage ICs

4.8.1 TTL
Clock speeds quoted are guaranteed.

Number	Description	Type	Clock MHz
Single JK			
7470	AND gate JK preset and clear	N	15↑
7471	AND gate JK preset	H	25↓
	—	L	2.5↓
7472	AND gate JK preset and clear	N	15⎍
	—	H	25⎍
	—	L	2.5⎍
74101	AND/OR gate JK preset	H	40↓
74102	AND gate JK preset	H	40↓
74110	AND gate JK preset clear	N	20⎍
Dual JK			
7473	Dual JK clear	N	15⎍
	—	H	25⎍
	—	L	2.5⎍
	—	LS	30↓
7476	Dual JK preset clear	N	15⎍
	—	H	25⎍
	—	LS	30↓
7478	Dual JK preset clear	H	25⎍
	—	L	2.5⎍
	—	LS	30↓
74103	Dual JK clear	H	40↓
74105	Dual JK preset clear	H	40↓
74107	Dual JK clear	N	15⎍
	—	LS	30↓
74108	Dual JK preset clear	H	40↓
74109	Dual JK preset clear	N	25↑
	—	LS	25↑
74111	Dual JK with lockout preset clear	N	20↓
74112	Dual JK preset clear	LS	30↓
	—	S	80↓
74113	Dual JK preset	LS	30↓
	—	S	80↓
74114	Dual JK preset clear	LS	30↓
	—	S	80↓

Quad JK

74276	Quad JK preset clear	N	35↓
74376	Quad JK clear	N	30↑
74379	Quad JK	LS	30↑

Hex and octal JK

74377	Octal JK	LS	30↑
74378	Hex JK	LS	30↑

D-types

7474	Dual D-type, preset clear	N	15↑
	—	H	35↑
	—	L	2.5↑
	—	LS	25↑
	—	S	75↑
74174	Hex D-type common clear	N	25↑
	—	LS	30↑
	—	S	75↑
74175	Quad D-type common clear	N	25↑
	—	LS	30↑
	—	S	75↑
74273	Octal D-type	N	30↑
	—	LS	30↑
74374	Octal D-type tristate	LS	30↑
	—	S	70↑
74377	Octal D-type with enable	LS	30↑
74378	Hex D-type with enable	LS	30↑
74379	Quad D-type with enable	LS	30↑
74388	Quad D-type tristate	LS	30↑

Hold/follow latches and SR flip-flops

7475	Quad latch	N, L, LS
74100	Eight-bit latch	N
74116	Dual four-bit latch	N
74118	Hex SR latch common reset	N
74119	Hex SR latch independent reset	N
74259	Eight-bit addressable latch	N, LS
74279	Quad NAND SR latch	N, LS
74373	Octal latch with tristate o/p	LS, S
74375	Quad latch	LS

4.8.2 CMOS
Clock speeds quoted are typical at 5 volts, guaranteed speed is roughly half the speed quoted below, 10 volts double and 15 volts triple.

Number	*Description*	*Clock MHz*
JK flip-flops		
4027	Dual JK preset clear	6↑
4095	AND gated JK preset clear	5↑
4096	AND gated JK preset clear	5↑
D-types		
4013	Dual D-type preset clear	12↑
4076	Quad D-type, tristate output	8↑
4174	(40174) Hex D-type	11↑
4175	(40175) Quad D-type	11↑
Hold/follow latches and SR flip-flops		
4042	Quad latch with selectable clock polarity	
4043	Quad NOR SR flip-flop	
4044	Quad NAND SR flip-flop	
4508	Dual quad latch, tristate output	
4597	Eight-bit latch tristate output	
4598	Eight-bit latch tristate output	
4599	Eight-bit addressable latch	

CHAPTER 5
Timers, Monostables and Oscillators

5.1 Introduction

In digital systems, there is often a need to generate a signal lasting a predetermined time. In industrial control, for example, it may be required to start an air purge fan and wait 10 seconds before igniting a gas burner. In computing, a delay of a few microseconds may be required before a signal received down a few metres of cable can be considered stable. The first part of this chapter will be concerned with devices giving time delays or pulses. These are variously known as delays, one-shots, strobes, monostables or (rather confusingly) flip-flops. There is, unfortunately, a certain amount of inconsistency in the definitions of timer circuits and functions.

Synchronous systems require some form of clock. The second part of this chapter will look at various oscillator circuits and ICs.

5.2 Delays on and off

Relay delays generally use pneumatic dashpots to slug the relay pick-up or drop-off. The equivalent logic circuits are known as delay on (fig. 5.1a) and delay off, (fig. 5.1b). The two circuits can be combined to give a delay on and off (fig. 5.1c). There is no generally accepted logic symbol for a delay, but the German DIN standard of fig. 5.1d is widely used.

Almost all timing circuits use an RC network to define the time. The simplest delay circuit is the delay on/off which can be implemented in CMOS or TTL form with the circuit of fig. 5.2. The output of gate a charges and discharges C via R. The voltage on C rises and falls exponentially, and the Schmitt trigger gate b gives a delayed output. With TTL, R must be a relatively low value, so the circuit can only be used where delays of less than a few tens of microseconds are required. There is no such restriction on CMOS, and delays of a few seconds are feasible. The timing accuracy is poor, and the period of the delay is approximately RC seconds.

The TTL and CMOS versions of a delay off circuit are shown in figs 5.3a and b respectively. The TTL version uses an open collector gate. The 1 to 0 transition at the output of gate a will be fast as C will be quickly discharged by the gate output transistor. The 0 to 1 transition will be an exponential rise as C charges via R. As the resistor only acts as a pull-up to the Schmitt input, it can have a value up to 10K.

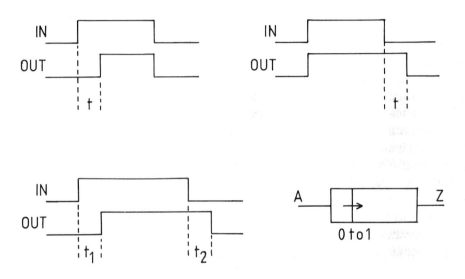

Fig. 5.1 Simple delay operations. (a) Delay 0 to 1 (delay on). (b) Delay 1 to 0 (delay off). (c) Delay 0 to 1 and 1 to 0. (d) DIN symbol for a delay.

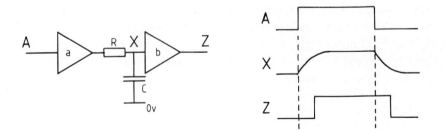

Fig. 5.2 Simple Schmitt trigger delay. (a) Simple delay circuit. (b) Operation.

The CMOS version uses a diode to discharge C. The timing capacitor is discharged via gate a and charged via R. Unlike TTL, a CMOS gate has a relatively high output impedance (typically 1K). To prevent an unacceptably long on-delay, the value of C should be as small as possible.

Both circuits will again have a delay of about RC seconds. Both circuits will exhibit a small on-delay as the timing capacitor discharges.

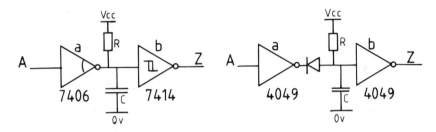

Fig. 5.3 Delay off (1 to 0) circuits. (a) TTL circuit. (b) CMOS circuit.

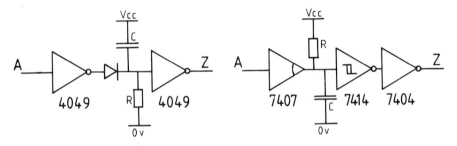

Fig. 5.4 Delay on (0 to 1) circuits. (a) CMOS circuit. (b) TTL circuit.

A CMOS delay-on circuit is implemented by reversing the diode and the RC arrangement as shown in fig. 5.4a. The capacitor is now discharged quickly on the 1 to 0 transition, and will charge slowly via R on the 0 to 1 transition. The TTL version uses the open collector 7407 non-inverting buffer. Unfortunately there is not a non-inverting Schmitt trigger in the TTL family, so two cascaded inverters need to be used to re-establish the correct polarity. Alternatively a non-inverting gate could be substituted for the two inverters, but this does incur a slight vulnerability to noise as the RC voltage passes through the indeterminate area between logic states.

5.3 Monostables (one-shots)

5.3.1 Introduction
A monostable produces a fixed-length pulse whenever a specified input condition occurs. A monostable responding to a 1 to 0 transition at its input would have a timing diagram similar to fig. 5.5a. In

all common monostables, the period is determined by an external resistor and capacitor.

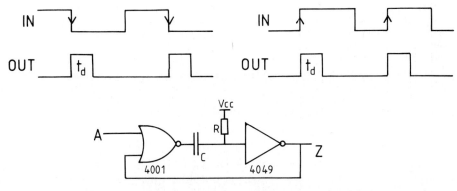

Fig. 5.5 Simple monostables. (a) Negative edge triggered. (b) Positive edge triggered. (c) CMOS positive edge triggered monostable.

5.3.2 Terminology

There are many variations on the basic monostables, and unfortunately there is no real standardisation of terms. The commonest monostable is the edge-triggered, which fires on a positive, or negative, edge. The circuit giving the waveform of fig. 5.5a is a negative edge triggered monostable, the circuit of fig. 5.5b is a positive edge triggered monostable. A simple CMOS positive edge triggered monostable is shown on fig. 5.5c.

An edge triggered monostable requires a triggering input from a logic gate (i.e. a relatively fast edge). Some monostables (e.g. the 74121) have a Schmitt trigger on one of the inputs. This allows the monostable to be fired by a slowly changing input as in fig. 5.6. Such inputs are said to be level triggered.

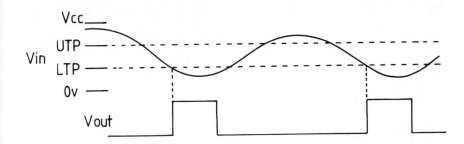

Fig. 5.6 Negative edge triggered monostable with Schmitt trigger input.

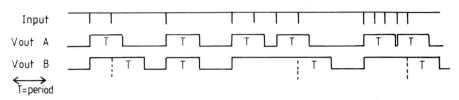

Fig. 5.7 Non-retriggerable (A) and retriggerable (B) monostables.

In simple edge triggered monostables, the output pulse period is unaffected by repetitive pulses. In fig. 5.7A, the output pulse period is the same regardless of the number of input pulses. A retriggerable monostable starts timing again with each input pulse, so the retriggerable monostable of fig. 5.7B extends the pulse period. The pulse width is defined from the last input trigger. Some monostables (such as the 74122) are inherently retriggerable. More versatile devices (such as CMOS 4047) can be used as a simple monostable or retriggerable monostable depending on its circuit connection.

Non-retriggerable monostables have a defined duty cycle. The period of a monostable is determined by the charging of a timing capacitor, through a timing resistor. When the voltage on the capacitor reaches some preset level the output pulse terminates. This action is summarised in fig. 5.8. When the pulse ends, the capacitor is in a charged state, and must be discharged by the monostable's internal circuit before the monostable can be used again. This discharge time (sometimes called the dead time) is related to the monostable period.

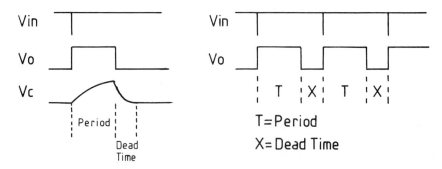

Fig. 5.8 Monostable dead time. (a) Cause of dead time. (b) Maximum duty cycle.

It follows that there is a maximum rate at which a particular monostable can be driven. This is usually defined as the Duty Cycle, which is:

$$\text{Duty cycle} = \frac{\text{On time}}{\text{Repetition rate}}$$

$$= \frac{T}{T + Td}$$

where T is the monostable period and Td the dead time.

The duty cycle depends on the value of the timing capacitor; the larger the capacitor value, the longer the dead time. Any given mono-stable period can be obtained with a wide range of resistors and capacitors. For example, all the combinations below give a time constant of 1 ms:

1K	1 μF
10K	0.1 μF
100K	0.01 μF

Choosing the smallest value capacitor will give the best duty cycle. Duty cycles are typically 60% to 90%. A monostable *can* be triggered during the dead time, but the resulting output will have a poorly defined period.

A device which is non-edge-triggered (and some retriggerable devices) can exhibit odd characteristics if the input pulse width is longer that the nominal monostable pulse width. A possible input waveform is shown in fig. 5.9. If this is fed to an edge triggered device, the expected output would be waveform (i). (This would be obtained from, say, a 74121.) Some devices would, however, give waveform (ii). These devices (of which the popular 555 is typical) cannot have an output pulse *shorter* than the input pulse unless some external edge triggered circuits (such as those in figs 4.11 to 4.13) are used.

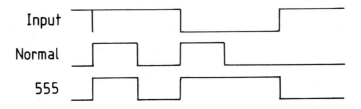

Input

Normal

555

Fig. 5.9 Effect of DC coupling used on 555 (and other) monostables.

5.3.3 Practical considerations

Monostables, by their very nature, turn short pulses into long pulses. They are probably the most noise-prone of logic elements. The precautions outlined in section 13.3 should be followed explicitly. In particular, input leads should be kept short and the supply decoupled local to the monostable IC. Monostables will fire on glitches from combinational logic circuits (see section 3.4).

Noise can enter via the timing component connections and cause random period variations. The timing components should be mounted as close as possible to the IC. For best noise immunity, a large capacitor and small resistor should be used. Unfortunately this conflicts with duty cycle requirements. Trimming of the period by an external trim pot should be avoided if possible, but if necessary should be done by a cermet trim pot mounted close to the monostable IC.

Long periods unfortunately require electrolytic capacitors. With the 74221 monostable, for example, the maximum value of timing resistor is 100K. A period of 1 second therefore requires a timing capacitor greater than 10 μF. Electrolytic capacitors have a large tolerance (typically -25%, $+50\%$), high leakage currents and are polarity sensitive. In addition, an electrolytic capacitor is voltage sensitive, and does not become a 'capacitor' until about 10 per cent of its rated voltage is reached.

If electrolytics are used with monostables, tantalum electrolytics should be chosen. These have stable values, relatively low leakage and small physical size. If a 5 volt supply is being used, 6 volt rated capacitors should be used; for a 12 to 15 volt supply, 15 volt rated capacitors should be used. Capacitors in monostables do not normally have to withstand more than two-thirds of the supply voltage. Polarity considerations are covered in different ways. Some monostables (such as the 74221) define a positive and negative terminal for the timing capacitors. Other monostables (such as the Fairchild 9600 series) require an external diode as in fig. 5.10 to prevent reverse biasing during discharge.

In general, electrolytics are less of a problem with CMOS monostables, because higher value timing resistors may be used (well in excess of 1 M for the 4528 for example). If very long periods are required (over about 10 seconds) the techniques outlined in section 5.5 should be used.

A monostable's period is generally very repeatable, although absolute accuracy is usually limited by the accuracy of the timing capacitor. A typical monostable will exhibit a drift of less than 0.5%

for 50°C change in temperature, and again about 0.5% for minimum to maximum supply voltage.

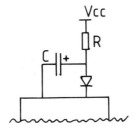

Fig. 5.10 Use of electrolytic timing capacitors.

Many monostables incorporate a clear input. This terminates the output pulse before the monostable's period. Its use should be tempered with care; some monostables (notably the 74221) will re-fire if the correct input conditions are present when the clear signal is removed. On some devices (such as the Fairchild 9600 series) the reset pulse resets the outputs, but does not discharge the timing capacitor, which times out normally. These devices cannot be fired again until after the normal timed period if a predictable pulse length is required. The clear input on all devices can also be used to inhibit the firing of the monostable.

5.4 Specific monostables

5.4.1 TTL 74122
There are four monostables in the TTL family; a retriggerable and non-retriggerable monostable in single and dual form. There is also a useful set of monostables in the Fairchild 9600 series. These are detailed in section 5.9. Typical of these is the 74122 shown in fig. 5.11. This is the single retriggerable monostable in the TTL family.

As can be seen, the 74122 has four trigger inputs; two low-going (A1/A2) and two high-going (B1/B2). The device can be fired on a positive edge (by B1 or B2 with both A inputs held low) or on a negative edge (by A1 or A2 with B inputs held high). The four inputs can be used to allow the monostable to fire if a specific logic combination occurs at the inputs. In Boolean algebra terms, the monostable will fire for (B1.B2.$\overline{\text{A1}}$.$\overline{\text{A2}}$). None of the inputs use Schmitt triggers, so normal TTL edge speeds must be provided. (Schmitt triggers inputs *are* provided on the LS version.)

The device is genuinely edge triggered, and the monostable only recognises edges. After firing the monostable, an input can remain in any state. The 74122 is a retriggerable monostable, but the retriggering can be inhibited by connecting the $\bar{Q}$ output to a B input if care is taken to ensure that the inputs are not in a 'true' state when the monostable times out. If this precaution is not followed, the 74122 may become an oscillator giving a rapid pulse train.

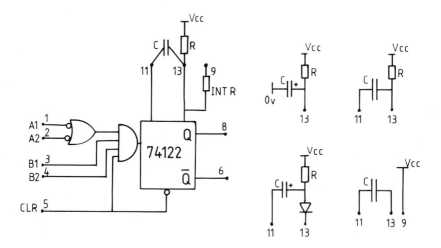

Fig. 5.11 The 74122 monostable. (a) Diagram. (b) Various RC connections.

The period is determined by an external resistor and capacitor connected as shown. The period is given by:

$$T = 0.4 \times R \times C \text{ seconds}$$

(This is a slight simplification, but is accurate enough for most purposes.) R can have any value between 5K and 50K (5K and 180K for the LS version). C can be any reasonable value, but stray capacitance will extend the period for low values, and leakage will extend the period in an unpredictable manner for large electrolytics. Possible RC connections are shown in fig. 5.11b. The timing resistor can be obtained from the Rint connection on pin 9. By linking pins 9 and 14 (Vcc) and connecting the timing capacitor as usual, a timing resistor of 5K is obtained.

A clear input is provided on pin 5. This terminates the monostable timing when pin 5 is taken low. If, however, a 'true' condition exists on the trigger inputs when the clear input goes high, the device will fire again.

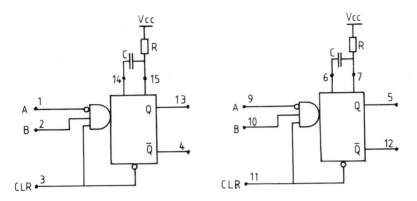

Fig. 5.12 The 74123 dual monostable.

The 74123 is a dual version of the 74122 with single positive and negative trigger inputs and clear input, shown on fig. 5.12. No internal timing resistor is provided.

5.4.2 CMOS 4528
The CMOS range of monostables consists of three dual monostables (4098, 4528, 4538) and three special-purpose devices (4047, 4541, 4753). The 4047 is a combined monostable/astable and is described later in this chapter. The 4541 and 4753 timers are described in chapter 14.

All the CMOS monostables are retriggerable, and typical is the 4528 shown in fig. 5.13. This is an improved version of the earlier 4098 and the two devices are pin-compatible.

Each monostable has two trigger inputs; one for positive edge triggering and one for negative edge triggering. A clear input is provided. The clear input is normally held high, but when taken low it terminates the output pulse. Unlike the 74121/2, the monostable will not fire again if the clear pulse is removed with a 'true' input condition present. Both trigger inputs incorporate Schmitt triggers, so the device can be used with slow edges.

The period is determined as usual by an external resistor and capacitor, and is given simply by:

$$T = RC \text{ seconds}$$

R can have any value from 10K to 10M, but leakage currents in capacitors make 1 M a more practical upper limit. There is no limit on the value of C, but the usual observations about stray capacitance for small values and leakage for large values apply. If an electrolytic is used, the positive lead should connect to the resistor.

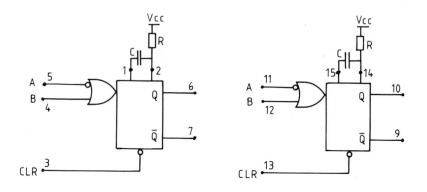

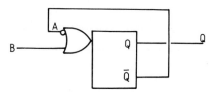

Fig. 5.13 4528 dual monostable. (a) Arrangement. (b) Non-retriggerable connection.

The 4528 provides two retriggerable monostables. They can be used as non-retriggerable monostables by linking $\bar{Q}$ to the negative input or Q to the positive input as shown in fig. 5.13b. A true condition must not exist on the triggering input at the termination of the timed period or unpredictable results may occur.

5.5 Long period timers

5.5.1 Introduction
The conventional monostables described in section 5.4 have a timed period of the order of RC seconds. For periods above a few seconds, large values of R and C are required with attendant problems from leakage, cost and physical size.

For periods above a few seconds it is usually easier to use a timer based on the principle of fig. 5.14. This uses a free-running oscillator and a counter to determine the period. Oscillators are described in section 5.7 and counters in chapter 7, but it is not necessary to have detailed knowledge of either to appreciate how fig. 5.14 performs.

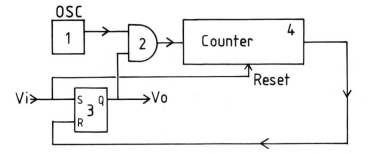

Fig. 5.14 Basic long period timer.

The oscillator 1 produces a free-running pulse train, which is normally blocked by gate 2. To start the monostable, a pulse is applied to the trigger input. This sets the flip-flop 3 and resets the counter 4. Gate 2 now passes pulses from the oscillator to the counter. When the counter reaches some predetermined value it resets flip-flop 3. The Q output of flip-flop 3 therefore goes high for a time given by:

$$T = N \times P \text{ seconds}$$

where N is the predetermined count for the counter and P the oscillator period. The period is accurate to one cycle of the oscillator.

For example, an accurate 100 second monostable could be constructed with a divide by 1000 counter and a 10 Hz oscillator (period 0.1 seconds). A 10 Hz oscillator can be made with reasonable value components.

A practical circuit is shown in fig. 5.15. IC1 is a 555 timer connected as an oscillator (see section 5.8.1). This is connected directly to a counter constructed of 7490 decade counters. As wired, the counter divides by 80 for two 7490s, 800 for three 7490s, 8000 for four and so on. In the practical circuit of fig. 5.15, the AND gate 2 of fig. 5.14 is superfluous as the counter is held in a reset state by gate 'c' when not required.

The flip-flop is constructed from half a 7400, (a, b). The two remaining 7400 gates are used as inverters to give the correct signal senses for the counter and flip-flop resets.

The circuit is triggered by a negative input pulse. It can be made retriggerable by linking pin 12 on gate 'c' to the input, or non-retriggerable by linking pin 12 to pin 13. If the retriggerable mode is used, each input pulse resets the counter to zero again.

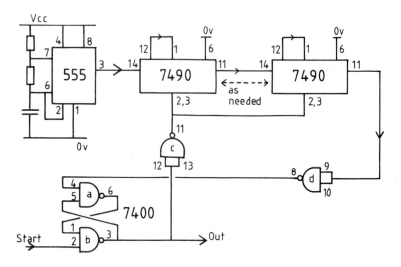

Fig. 5.15 Practical long period timer.

In a clocked synchronous system, it is often convenient to use the system clock. As this is usually of the order of 1 MHz a very large counter is required. The CMOS family contains many useful counters such as the seven-bit 4024, 12-bit 4040, 14-bit 4060 and 24-bit 4521 which can convert from a system clock to a few Hertz with one IC package. There are no equivalent devices in the TTL family, but CMOS and TTL can, of course, be intermixed as explained in section 13.4. If a crystal-controlled system clock is used, very accurate time delays can be obtained.

5.5.2 Ferranti ZN1034 precision timer
This useful IC is effectively fig. 5.14 in a single device. It is shown in block diagram form in fig. 5.16. As can be seen, it incorporates an oscillator, 12-bit (divide by 4096) counter and control logic. The IC itself operates on 5 volts, but incorporates a *shunt* regulator to allow it to be used with other supplies.

The oscillator frequency is determined by a timing resistor (in the range 5K to 5M) and a capacitor connected as shown (Rt, Ct) and also by a calibration resistor Rk connected between pins 11 and 12. The device has an internal 100K calibration resistance, so the total value of calibration resistance Rcal is Rk + 100K. The monostable period is given by:

$$T = K \times 4095 \times Rt \times Ct \text{ seconds}$$

where K is a calibration constant determined by Rcal as follows:

Rcal	*K*	
100K	0.67	(pins 11, 12 linked)
125K	0.8	
175K	1.0	
225K	1.2	
300K	1.4	(maximum value)

Normally a 47K trim pot is connected between pins 11 and 12, and a value of 0.8 assumed for K. The trim pot can then be used to set an accurate time.

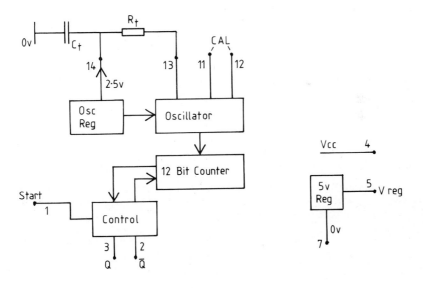

Fig. 5.16 The ZN1034 long period timer.

Long periods can be obtained with reasonable value components. For example:

Rt	Ct	Rcal	Period
100K	1 μF	100K	5 mins
1M2	1 μF	300K	2.5 hours
2M2	100 μF	100K	1 week

In the latter case, a low leakage capacitor should be used. The minimum suggested period is 50 ms.

The device is edge triggered, and can be fired by taking pin 1 low. If pin 1 is tied to 0 V, the device will fire once on power up. It is not retriggerable. True and complement outputs are available, which will sink or source 25 mA.

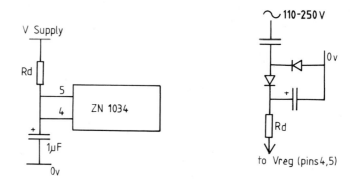

Fig. 5.17 Use of ZN1034 on high voltage supplies. (a) Use of internal regulator. (b) Lossless dropper.

If the shunt regulator is used, a series resistor should be used from the supply as shown in fig. 5.17a. The required value of Rd (in kilohms) is given by:

$$Rd = \frac{(V-5)}{10}$$

where V is the supply voltage. If more than a few milliamps is being drawn from the ZN1034 outputs in the high state the value of Rd should be reduced accordingly. It is probably of little use to the average logic designer, but it is worth noting that the ZN1034 can be driven directly off 240 volts AC using the lossless dropper circuit of fig. 5.17b.

5.5.3 CMOS 4541 programmable timer
The 4541 programmable timer is the CMOS equivalent of the ZN1034, and is shown in the block diagram on fig. 5.18. An RC

controlled oscillator is fed to a binary counter whose length can be selected by control inputs A0, A1. The device can be used as a frequency divider or a monostable.

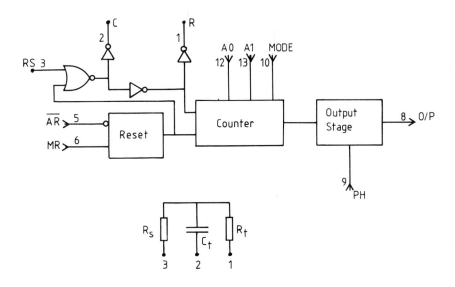

Fig. 5.18 The 4541 programmable counter.

The oscillator frequency is determined by Rt and Ct, and is given by:

$$f = \frac{1}{2.3 \times Rt \times Ct}$$

Rs should be chosen to be about twice Rt. The value of Rs does not affect the frequency. Rt can have any value from 10K to 1M. Ct can have any reasonable value, but the device is not really suitable for use with electrolytic capacitors. If required, the internal oscillator need not be used, and an external clock can be connected to the RS pin.

The oscillator output is fed to a binary counter whose length is selected by control inputs A0, A1 as follows:

A0	A1	Stages	Division
1	0	8	256
0	1	10	1024
0	0	13	8192
1	1	16	65 536

The mode input determines whether the device acts as a monostable (Mode = 0) or as a frequency divider (Mode = 1).

If the monostable mode is selected, the device can be triggered on power up or via a trigger input. Auto triggering (called auto reset) is enabled with a 0 on the $\overline{AR}$ input. Manual triggering is activated by a 0−1−0 pulse on the MR input.

Only a single output, 0, is provided but this can be in true or complement form as selected by the PH input. With PH = 0, the output is a 1−0−1 pulse (complement output in usual monostable notation). With PH = 1, a true 0−1−0 output pulse is obtained.

The 4541 is a very useful device where long time delays are required, and (like the ZN1034) requires reasonable value components. For example, using the 16-stage counter a 1 minute delay can be obtained with Rt = 470K and Ct = 1 nF.

5.6 Strobes

A strobe is a short pulse whose length need not be accurately defined. They are generally used to produce a pulse for setting or resetting a flip-flop, gating data and similar applications. A strobe can be represented by fig. 5.19a. Obvious variations are triggering on positive or negative or both edges, and inverted pulses. Most strobe circuits are based on the RC differentiator of fig. 5.19b which produces a poorly shaped pulse which can be sharpened by a Schmitt trigger. A diode across R is needed to clip the negative pulse.

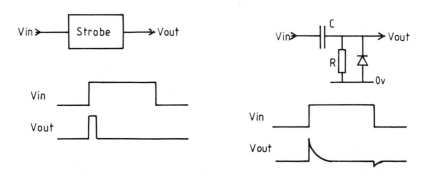

Fig. 5.19 Basics of strobe circuits. (a) Strobe circuit. (b) RC differentiator.

Practical TTL circuits are shown in fig. 5.20a (negative edge trigger) and fig. 5.20b (positive edge trigger). The negative edge trigger circuit requires bias resistors to give an input to the 7414 corresponding to a logic 1. If the input were simply tied to the 5 volt rail, a normal TTL 1 to 0 transition would not take the gate input low enough. Note the low value of the resistor in fig. 5.20b. The output pulse width is poorly defined, but is the order of 0.5RC seconds. Note that the designer can only choose the value of C.

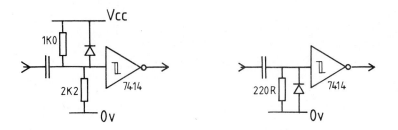

Fig. 5.20 TTL (inverting) strobe circuits. (a) Negative edge. (b) Positive edge.

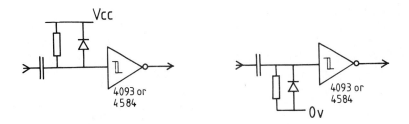

Fig. 5.21 CMOS (inverting) strobe circuits. (a) Negative edge. (b) Positive edge.

The circuits shown produce an inverted pulse. A true pulse can be obtained by adding an inverter or replacing the 7414 with a non-inverting buffer or gate if the slight degradation of the edge speed can be tolerated.

The equivalent CMOS circuits are shown in fig. 5.21. The circuit will work with most B series gates or inverters. The output pulse width is again about 0.5RC seconds, but with the CMOS circuits the designer can choose both R and C.

Some precautions should be observed in using these circuits. The input pulse width must exceed the strobe pulse width by a factor of five, or the capacitor will not discharge fully. If this is not possible, the circuit of fig. 5.5c may be used. It should also be noted that there may be some degradation of the input signal edges, caused by the clamping action of the protection diodes.

An alternative circuit based on the delays of section 5.2 is shown on fig. 5.22. This gates a delayed inverted version of the signal with itself to produce a pulse. The circuit can provide a pulse on the positive edge, negative edge, or both by suitable choice of buffers/ inverters in the delay part of the circuit.

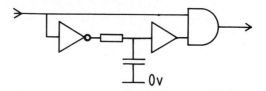

Fig. 5.22 Alternative strobe circuit.

5.7 Oscillators and clocks

5.7.1 Introduction

Oscillators, or clocks as they are more commonly known, are found in timers, synchronous systems (see sections 4.5 and 7.4), multiplexed displays (see section 9.5) and many other applications. This chapter examines several possible oscillator circuits.

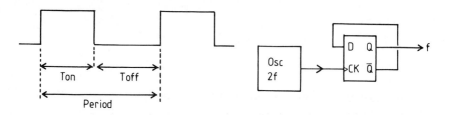

Fig. 5.23 Oscillator basics. (a) Definitions of terms. (b) 50% duty cycle.

The output from a clock will be a square wave pulse train as in fig. 5.23a. The on time is defined as the time the waveform spends in the 1 state, the off time is defined as the time in the 0 state. The duty cycle is defined as:

$$\text{Duty cycle} = \frac{\text{Ton}}{\text{Period}} \text{ (expressed as a percentage)}$$

Sometimes the term mark/space ratio is used, which is defined as Ton/Toff.

An exact 50% duty cycle (Ton = Toff) is required for some applications (driving liquid crystal displays for example). This is difficult to provide with any clock circuit, but can be obtained by using an oscillator running at *twice* the required frequency followed by a toggle flip-flop (toggle flip-flops are described in section 7.2).

Figure 5.23b shows a D-type flip-flop used as a toggle flip-flop and giving an exact 50% duty cycle from an oscillator output. It should be noted that any 'jitter' in the oscillator output due to, say, power supply noise, will cause the duty cycle to vary transiently from the 50% duty cycle.

5.7.2 Simple oscillators

The simplest oscillators can be constructed from Schmitt triggers. The circuit shown in fig. 5.24 can be used with TTL (7414) or CMOS (4584). The timing capacitor charges and discharges via the timing resistor, giving an exponential voltage at point A, which cycles between the upper and lower trigger points.

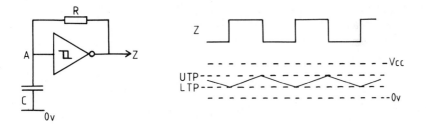

Fig. 5.24 Schmitt trigger oscillator. (a) Circuit. (b) Waveforms.

The CMOS version will work with any reasonable values of R and C, even with electrolytics. The oscillator frequency is approximately given by:

$$f = \frac{3}{RC} \text{ Hz}$$

The TTL version requires a value of 330R for R (to give a defined 0 state as explained in section 2.6). The frequency is determined by the value of C, and is approximately given by:

$$f = \frac{1}{RC} \text{ Hz}$$

The above formula may be in error by as much as 50%, and the duty cycle will probably not be 50% (particularly for the TTL version with its less well-defined levels). The circuit does give a very cheap oscillator which can be made to operate from about 10 Hz to 1 MHz.

In CMOS only, an improved oscillator can be made from two

inverters (which need not be Schmitt triggers) as shown in fig. 5.25. Suppose point b has just gone to a 1. Point a must just have gone to a 0, and Ct will start to charge via Rt, giving a 1 decaying exponentially at point c. This maintains the 0 at point a via gate 2, until the voltage at c reaches 0.5 Vcc when gate 1 switches. Point a goes to a 1, point b goes to a 0, and point c goes to a 0 but starts rising exponentially as c charges in the opposite polarity. The 0 at point c maintains the 1 at point a until the voltage at point c again reaches 0.5 Vcc when gate 1 switches again. Point a goes to a 0, and point b to a 1 again. The cycle now repeats and the circuit oscillates at a period determined by Rt and Ct.

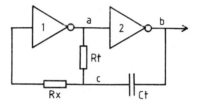

Fig. 5.25 CMOS oscillator.

Rx is included to prevent degradation of the capacitor voltage by the input protection diodes of gate 1. Rx should be chosen to be 5 or 10 times the value of Rt; it does not affect the oscillation frequency.

Assuming a switching point for gate 1 of 0.5 Vcc, the period is 1.4RC seconds, giving a frequency of:

$$f = \frac{0.7}{RC} \text{ Hz}$$

The duty cycle depends on the switching point, but is nearly 50%.

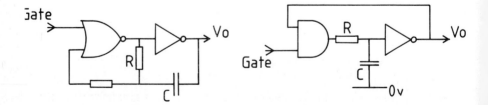

Fig. 5.26 Gated oscillators. (a) CMOS oscillator. (b) TTL or CMOS oscillator.

A gated oscillator can be formed by having a free-running oscillator followed by a gate, or by one of the gated oscillators shown on fig. 5.26a (for CMOS) and fig. 5.26b (for TTL). The latter have the advantage of a predictable starting pattern.

5.7.3 Clock circuits

The simple circuits of section 5.7.2, whilst cheap and easy to construct, are not very stable and will yield different frequencies with supposedly identical ICs. Where a more predictable or stable frequency is required, one of the circuits described in this section should be used.

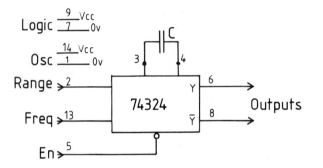

Fig. 5.27 The 74LS324 voltage-controlled oscillator.

Typical of these clock generator ICs is the TTL 74LS324 voltage controlled oscillator shown in fig. 5.27. This can be used to generate a stable clock signal over the range 0.1 Hz to 30 MHz.

The frequency is set by a capacitor connected between pins 3 and 4, and can be adjusted by means of an analog voltage in the range 0 to 5 volts applied to pins 2 and 13. These inputs are known respectively as Range and Frequency Control, and can be adjusted independently, giving an adjusted range of about 3:1. With both inputs at 2 volts, the output frequency is given by:

$$f = \frac{10^{-4}}{C} \text{ Hz where C is in farads}$$

At high frequencies, C will be measured in picofarads, and the effect of stray capacitance will be significant.

True and complement outputs are available. Separate supply pins are provided for the oscillator and output logic to minimise noise problems. Normally they would run on the same power supply but have separate paths back to the supply to prevent logic-induced supply noise affecting the oscillator. An enable pin turns the output signals on (low for enable). Internal synchronisation ensures that no 'half' pulse occurs on enabling or disabling the output.

The 74LS324 is one of a range of VCOs in the TTL family (74LS324-74LS327). The other members include dual versions with various options such as enable. There is no equivalent CMOS version, but a VCO is included in the 4046 phase-locked loop IC.

Edge triggered monostables can be coupled together to form a clock circuit. This approach is particularly attractive where a specified duty cycle has to be provided, as each monostable can be adjusted independently. Figure 5.28 shows a clock using both halves of a dual monostable such as the 74123 or 4098.

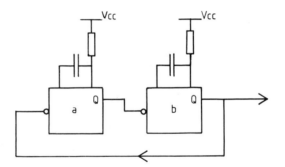

Fig. 5.28 Oscillator using edge triggered monostables.

5.7.4 Crystal oscillators

If a high-accuracy, very stable, clock is required a crystal oscillator should be used. With care, accuracies of better than 10 ppm can be achieved.

The 74LS324 family described earlier are designed to function as crystal oscillators when the timing capacitor is replaced by a crystal. A stable oscillator can therefore be constructed with one IC and a crystal.

Simpler circuits can be constructed from simple inverters although the stability is not as good as specific ICs such as the 74LS324. Figure 5.29a shows a simple TTL oscillator and fig. 5.29b a CMOS version. It is also possible to construct a simple transistor oscillator followed by a buffer as in fig. 5.29c.

All these crystal circuits operate at the fundamental crystal frequency. Many RF crystals operate at some harmonic. If an RF crystal is used in one of the above circuits, it is possible that the resulting frequency will be some sub-multiple of the crystal's marked frequency.

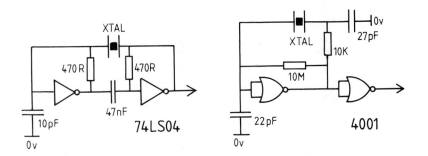

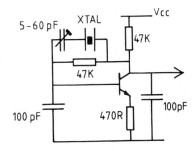

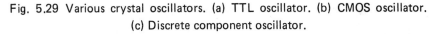

Fig. 5.29 Various crystal oscillators. (a) TTL oscillator. (b) CMOS oscillator. (c) Discrete component oscillator.

A crystal's frequency can be trimmed slightly by using a series or parallel capacitor. Figure 5.29c uses a small adjustable trimmer. In timing applications, the following crystal frequencies are commonly used:

Output frequency	Binary stages	Crystal frequency
50 Hz	16	3.276800 MHz
1 Hz	22	4.194304 MHz

For very critical applications, temperature controlled crystal oscillator modules can be obtained. These contain a small heating element and temperature control circuits to keep the frequency independent of ambient temperature.

Clock generator ICs for microprocessors (such as the 8080) use crystal oscillators, and can, of course, be used in circuits outside their own family.

5.8 Miscellaneous devices

5.8.1 The 555 timer

The 555 is one of the most versatile timers available, and can be made to perform as a monostable or an astable. It is shown in block diagram form in fig. 5.30. The circuit contains two comparators which are triggered at one-third Vcc (Trigger input) and two-thirds Vcc (Threshold input). These set and reset a memory flip-flop which controls an output stage and a discharge transistor (whose use we will shortly discuss).

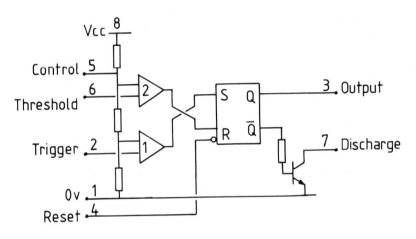

Fig. 5.30 The 555 timer.

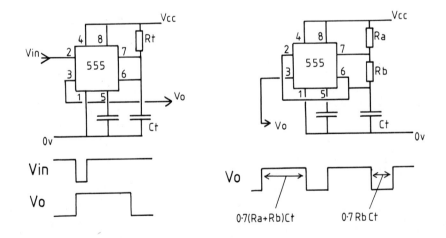

Fig. 5.31 Basic 555 circuits. (a) Monostable circuit. (b) Astable circuit.

To construct a monostable we add a timing resistor and capacitor as in fig. 5.31a. A low-going pulse to the trigger input sets the memory via comparator 1. The discharge transistor turns off, and Ct starts to charge via Rt. When the voltage at pin 6 reaches two-thirds of Vcc, comparator 2 resets the memory. This discharge transistor turns on again to discharge Ct ready for the next operation. The output is a high-going pulse of period:

$$T = 1.1 \times R \times C \text{ seconds}$$

The 555 is a DC coupled device, and as described in section 5.3.2, the input pulse must be shorter than the monostable period. If this is not possible, an RC edge trigger circuit must be added.

An astable is constructed as shown in fig. 5.31b, with two resistors and a capacitor. Assume the memory has just been set. Capacitor Ct will be charging via (Ra + Rb). When the voltage on Ct reaches two-thirds of Vcc, comparator 2 will trigger and reset the memory. Ct now discharges via Rb. When the voltage on Ct reaches one-third of Vcc, comparator 1 (via the trigger input) sets the memory again. Ct starts to charge repeating the cycle. The output is a square wave, and the voltage on Ct rises and falls between one-third and two-thirds of Vcc.

The output is high for a period:

$$Th = 0.7 (Ra + Rb) Ct \text{ seconds}$$

and low for a period:

$$Tl = 0.7 Rb Ct \text{ seconds}$$

It follows that the duty cycle cannot be less than 50% in the simple astable.

The 555 is a very versatile device, and it is possible to use it in many ingenious circuits. The control voltage on pin 5 (which is normally decoupled with a 0.01 μF capacitor) can be used as an input to adjust the frequency, and the addition of diodes to control the charge and discharge of Ct allows duty cycle control. Such techniques are rather outside the scope of this book however.

The 555 is also available in dual form (the 556) and quadruple form (the 557 and 558). A low power CMOS version is also available.

5.8.2 CMOS 4047 astable/monostable

This very versatile member of the CMOS family can be used as an astable or an edge triggered monostable with retriggerable option. Logic inputs determine the mode of operation, so external circuits can select and change the device characteristics.

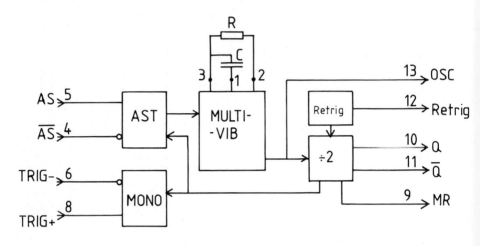

Fig. 5.32 The 4047 astable/monostable.

The 4047 is shown in block diagram form in fig. 5.32. In all modes, timing is determined by a single resistor and capacitor. The mode is selected by the five inputs as follows:

Mode	Astable	$\overline{\text{Astable}}$	Positive trigger	Negative trigger	Retrigger
Free-running astable	1	1	0	1	0
Positive gating	gate	1	0	1	0
Negative gating	0	gate	0	1	0
Positive edge trigger monostable	0	1	trigger	0	0
Negative edge trigger monostable	0	1	1	trigger	0
Retriggerable monostable	0	1	trigger	0	trigger

In the retriggerable mode the positive edge trigger and retrigger inputs are linked. The device retriggers on positive edges.

In the astable mode, an exact 50% duty cycle is provided by the output toggle stage. The actual oscillator output (at twice the output frequency) is available on pin 13, but this does not have a predictable duty cycle. It can, however, be used for long delay applications as described in section 5.5. The oscillator output can be connected to an external counter whose final output is, in turn, connected back to pin 4 to stop the oscillator.

In the astable mode, the period is 2.2RC seconds. In the monostable mode, the pulse width is 2.5RC seconds.

A reset facility is provided on pin 9. A 1 resets the output (and inhibits further triggering).

5.8.3 4702 Baud rate generator

There are many specialised ICs which are basically timers or astables. Typical of these is the Fairchild 4702 Baud rate generator (which despite its number is *not* a CMOS device). This useful IC was designed to produce clock pulses for the common data rates used in serial transmission.

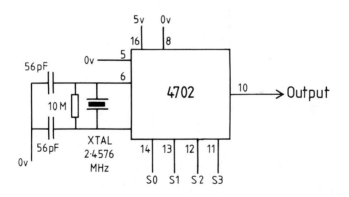

Fig. 5.33 Baud rate generator.

The device, shown in fig. 5.33, contains a crystal oscillator which operates at 2.4576 MHz. This is divided down by internal counters to give the correct clock rates for the various Baud rates. The clock frequencies are 16 times the Baud rate. The 110 Baud output is, for example, 1760 Hz.

S0	S1	S2	S3	Baud Rate	Freq (Hz)
1	1	1	1	110	1 760
0	1	1	1	150	2 400
1	0	1	1	300	4 800
0	1	1	0	600	9 600
1	1	0	1	1200	19 200
0	1	0	1	1800	28 800
1	1	1	0	2400	38 400
1	0	0	1	4800	76 800
0	0	0	1	9600	153 600

Other combinations are unused or produce duplications. With the use of an external multiplexer such as the 74LS259, eight simultaneous Baud rates can be provided.

5.9 Monostable/timer and oscillator ICs

5.9.1 TTL

Monostables		*Availability*
74121	Monostable with Schmitt input	N, L
74122	Retriggerable with clear	N, L, LS
74123	Dual retriggerable with clear	N, L, LS
74221	Dual monostable with Schmitt input	N, LS
74422	Retriggerable with clear	LS
74423	Dual retriggerable with clear	LS

Oscillators		
74124	Dual voltage controlled oscillator	LS, S
74297	Digital phase locked loop	LS
74320/321	Crystal oscillator	LS
74324-74327	Voltage controlled oscillators	LS
74624-74629	Voltage controlled oscillators	LS

5.9.2 CMOS

Monostables and timers
4047   Monostable/astable multivibrator
4098   Dual monostable
4528   Dual monostable with clear
4536   Programmable timer

Monostables and timers

4538	Dual monostable
4541	Programmable timer
4566	Industrial timebase
4753	Universal timer

Oscillators

4046	Phase locked loop

5.9.3 Other devices

Monostables/timers

555	Universal timer, monostable/astable (also available in CMOS form)
556	Dual 555
557/8	Quad 555
2240/7240/7250	Timer/counters similar to fig. 5.14
ZN 1034	Timer/counter similar to fig. 5.14
ICM 7242	Timer/counter similar to fig. 5.14

Oscillators

M706B1	Crystal oscillator with divider to give 50 Hz output
4702	Baud rate generator (with crystal oscillator)
ICL 8038	Waveform generator
MM 5368	Crystal oscillator with divider to give 50/60 Hz, 10 Hz, 1 Hz

CHAPTER 6
Binary Arithmetic

6.1 Introduction

In previous chapters logic signals have been assumed to represent an event. Printer Ready, Low Oil Level, Select Mode Switch are the sorts of signals that have been represented in digital 1 or 0 form. Digital signals can also be used to represent numbers, and digital circuits can be built to manipulate these numbers and perform the arithmetical operations of addition, subtraction, multiplication and division.

This chapter will be concerned solely with the theory behind binary numbers and binary arithmetic. Chapter 7 will convert this theory into practice.

6.2 Number systems, bases and binary

We are so used to our conventional decimal number system that it is usually assumed that it is in some way 'natural'. Normal day-to-day calculations are executed in multiples of ten. For example, the decimal number 4057 means:

	4 thousands	$= 4 \times 10 \times 10 \times 10$
plus	0 hundreds	$= 0 \times 10 \times 10$
plus	5 tens	$= 5 \times 10$
plus	7 units	$= 7$

Our decimal counting system goes:
1, 2, 3, 4, 5, 6, 7, 8, 9, 10, 11 . . . . , 97, 98, 99, 100, 101 . . .

Each position in a decimal number represents a power of ten. The square of ten is a hundred; the cube of ten is a thousand. In mathematical terms, conventional arithmetic is done to a base of ten.

The reason that day-to-day calculations are done to a base of ten is simply because human beings have ten fingers, and early arithmetic was done literally by hand. If we had eight fingers we would prob-ably count to base 8.

A number system can be constructed to any base. Table 6.1 shows comparison between counting in various bases. Of particular interest to us are the bases 8 (called octal), 16 (called hex for hexadecimal) and 2 (called binary). Other bases are used in everyday life; inches count to base 12; days of the week to base 7.

Table 6.1 Counting to various bases

Decimal	Binary	Octal	Hex
0	00000	0	0
1	00001	1	1
2	00010	2	2
3	00011	3	3
4	00100	4	4
5	00101	5	5
6	00110	6	6
7	00111	7	7
8	01000	10	8
9	01001	11	9
10	01010	12	A
11	01011	13	B
12	01100	14	C
13	01101	15	D
14	01110	16	E
15	01111	17	F
16	10000	20	10
17	10001	21	11
18	10010	22	12
19	10011	23	13
20	10100	24	14
etc	etc	etc	etc

A number system to base 2 needs only two symbols; 0 and 1. Each position in a binary number represents a power of two in the same way that each position in a decimal number represents a power of ten.

For example 1011010 is evaluated:

$$1 \times 2 \times 2 \times 2 \times 2 \times 2 \times 2 \quad = 64$$
$$0 \times 2 \times 2 \times 2 \times 2 \times 2 \quad = 0$$
$$1 \times 2 \times 2 \times 2 \times 2 \quad = 16$$
$$1 \times 2 \times 2 \times 2 \quad = 8$$
$$0 \times 2 \times 2 \quad = 0$$
$$1 \times 2 \quad = 2$$
$$0 \times 1 \quad = 0$$

Decimal	90

Binary numbers are written in the same format as decimal numbers, with the least significant digit to the right. A binary digit (i.e. a single 1 or 0) is usually called a 'bit'.

Four bits (called a nibble) can represent any number in the range 0-15. Eight bits (called a byte) can represent any number in the range 0-255. Sixteen bits (sometimes called a word) can represent any number in the range 0-65 535. In general, n bits can represent any number between 0 and $2^n - 1$ (for comparison, n decimal digits can represent 0 to $10^n - 1$).

The use of binary to represent numbers offers many attractions to the designer. Any other base requires some form of analog representation and noise-prone voltage comparators. Binary uses just two states, 0 and 1, which can be represented by two voltages. The resulting circuits are consequently simple, cheap and noise-tolerant. The only real disadvantage in using binary comes at the interface between the binary circuit and the outside world where decimal/binary conversion is performed on inputs and binary/decimal conversion on outputs.

Binary to decimal conversion is performed by simply summing powers of two as shown in the example above. Two more examples are given below:

(a) 1101. This is evaluated:

$$1 \times 8 \quad = \quad 8$$
$$1 \times 4 \quad = \quad 4$$
$$0 \times 2 \quad = \quad 0$$
$$1 \times 1 \quad = \quad 1$$

	13

Binary 1101 is decimal 13

(b) 1010 0110

$$1 \times 128 \quad = \quad 128$$
$$0 \times 64 \quad = \quad 0$$
$$1 \times 32 \quad = \quad 32$$
$$0 \times 16 \quad = \quad 0$$

```
0 × 8   =   0
1 × 4   =   4
1 × 2   =   2
0 × 1   =   0
              166
```
Binary 1010 0110 is decimal 166

   Decimal to binary conversion can be achieved in many ways, but the simplest is to perform successive divisions by two, noting the remainders. Reading the remainders from top (LSB) to bottom (MSB) gives the equivalent binary number. This is, in fact, simpler than the description implies, and is best illustrated by a few examples:

(a) Decimal 23

```
23
11          r 1 least significant bit
 5          r 1
 2          r 1
 1          r 0
 0          r 1 most significant bit
```
Decimal 23 is binary 10111

We can check this by noting that 10111 converted to decimal is 16 + 4 + 2 + 1 = 23.

(b) Decimal 109

```
109
 54          r 1
 27          r 0
 13          r 1
  6          r 1
  3          r 0
  1          r 1
  0          r 1
```
Decimal 109 is binary 110 1101

(c) Decimal 12

```
12
 6          r 0
 3          r 0
 1          r 1
 0          r 1
```
Decimal 12 is binary 1100

Binary/decimal and decimal/binary conversion is very straight-forward once it has been practised a few times. The reader is advised to attempt conversions of various numbers until proficiency is attained.

6.3 Octal and hexadecimal representation

A straight binary number such as 1101 0011 means very little even to the experienced designer. It can be converted to decimal (giving the result 211) but the operation is a bit tedious and often gives no indication of the binary pattern.

A 'shorthand' way of representing a binary number is often useful. There are two ways of achieving this. The first represents the number in base 8 form (octal), the second represents the number in base 16 form (hex).

Octal digits can have values from 0 to 7. These have the binary equivalents:

Octal	Binary
0	000
1	001
2	010
3	011
4	100
5	101
6	110
7	111

To convert a binary number to octal representation, the binary number is grouped in three-bit sets from the least significant bit. The octal equivalent of each group is then noted underneath. Like most binary operations, this is best illustrated by examples:

(a) 11010011

Grouped	11	010	011
Octal	3	2	3

The octal representation is 323

(b) 101010111001

Grouped	101	010	111	001
Octal	5	2	7	1

The octal representation is 5271

(c) 1110010101
    Grouped   1   110   010   101
    Octal   1   6   2   5
    The octal representation is 1625

Conversion from octal to binary is equally straightforward. The binary equivalent of each octal digit is simply written underneath. Again, examples are the best explanation:

(a) Octal 4057
    Octal   4   0   5   7
    Binary   100   000   101   111
    The equivalent binary number is 100000101111

(b) Octal 3265
    Octal   3   2   6   5
    Binary   11   010   110   101
    The equivalent binary number is 11010110101

Hex representation is built around base 16. To construct a number system to base 16, an additional six symbols are needed to represent decimal 10, 11, 12, 13, 14 and 15. It is usual to use the letters A, B, C, D, E, F.

Counting from 0 to 16 in decimal, hex and binary we get:

Decimal	Hex	Binary
0	0	0000
1	1	0001
2	2	0010
3	3	0011
4	4	0100
5	5	0101
6	6	0110
7	7	0111
8	8	1000
9	9	1001
10	A	1010
11	B	1011
12	C	1100
13	D	1101
14	E	1110
15	F	1111
16	10	10000	etc.

To represent a binary number in hex form, it is split into four-bit groups (or nibbles). The hex representation for each group can then be written down as shown in these examples:

(a) 1011100

Grouped	101	1100
Hex	5	C

The hex representation is 5C

(b) 11110101

Grouped	1111	0101
Hex	F	5

The hex representation is F5.

(c) 10101001101011

Grouped	10	1010	0110	1011
Hex	2	A	6	B

The hex representation is 2A6B

The conversion from hex to binary is performed in a similar manner to the conversion from octal to binary. The (four-bit) binary equivalent of each hex digit is simply written down in order. For example:

(a) 37C2

Hex	3	7	C	2
Binary	11	0111	1100	0010

The binary equivalent is 11011111000010

(b) D4

Hex	D	4
Binary	1101	0100

The binary equivalent is 11010100

Hex and octal representations are often used in microcomputer instructions. It is considerably easier to remember that C3 means Unconditional Jump in a microprocessor's instruction set than it is to remember the equivalent eight-bit binary number 1100 0011.

Confusion can arise when different number systems are intermixed. It is usual to write hex and octal numbers with a suffix showing the base being used:

(97)H is 97 in base 16 (i.e. a hex number)
(177)8 is 177 in base 8 (i.e. octal)
275 is decimal 275 (no suffix)

(97)H is, of course, decimal 151 and (177)8 is decimal 127.

6.4 Addition of binary numbers

Before binary addition is described, it is useful to talk through a simple decimal addition:

$$
\begin{array}{r}
345 \\
+\ 272 \\
\hline
617 \\
\hline
\end{array}
$$

Methods differ according to schools' teaching practices, but most people verbalise it:

'Five plus two is seven, no carry. Four plus seven is eleven, that's one down and carry one. Three plus two plus the carry is six with no carry. Result six, one, seven; that's six hundred and seventeen'.

Addition in decimal is taken a digit at a time starting with the least significant. At each stage we consider three 'inputs': the two digits and a possible carry input from the preceding stage. Each stage has two outputs: a sum and a possible carry to the succeeding stage. One stage could be represented by fig. 6.1. Identical stages could be combined as in fig. 6.2 to give an adder of any required length.

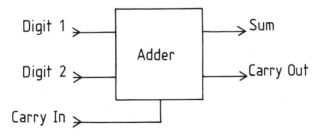

Fig. 6.1 Single digit full adder.

Addition of more than two numbers is not significantly different, as the operation can be broken down into several stages, each of which involves two additions. For example, $325 + 627 + 124 + 902$ can be performed in three stages:

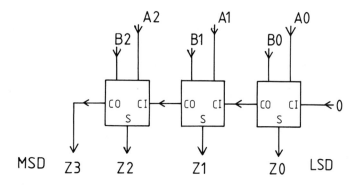

Fig. 6.2 Three-digit adder constructed from three single-digit adders. Two three-
digit numbers A, B are added to give a four-digit result, Z.

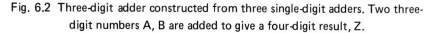

$$325 + 627 = 952$$
$$+ 124 = 1076$$
$$+ 902 = 1978$$

1978 being the required answer.

Binary addition is, in principle, identical except that the digits can only have values 0 or 1. At each stage there are eight possible input combinations (two inputs and a carry input from the preceding stages). These produce a sum and carry output. All possible combinations can be tabulated in an eight-line truth table:

Inputs			Outputs	
Carry	B	A	Sum	Carry
0	0	0	0	0
0	0	1	1	0
0	1	0	1	0
0	1	1	0	1
1	0	0	1	0
1	0	1	0	1
1	1	0	0	1
1	1	1	1	1

Circuits to perform addition using this truth table are described in chapter 8. Using the truth table, some examples of binary addition are:

(a)
$$
\begin{array}{l}
1\ 0\ 1\ 1\ 0\ 1\ 0 \\
\underline{0\ 1\ 0\ 1\ 0\ 1\ 1}
\end{array}
$$

Sum   1 0 0 0 0 1 0 1

Carry   1 1 1 1 0 1 0

(b)
$$
\begin{array}{l}
1\ 1\ 0\ 1\ 1\ 0 \\
\underline{1\ 0\ 0\ 0\ 1\ 1}
\end{array}
$$

Sum   1 0 1 1 0 0 1

Carry   1 0 0 1 1 0

(c)
$$
\begin{array}{l}
1\ 0\ 1\ 1\ 0\ 1\ 0 \\
\underline{\qquad\ \ 1\ 1\ 1}
\end{array}
$$

Sum   1 1 0 0 0 0 1

Carry   0 1 1 1 1 0

(d)
$$
\begin{array}{l}
1\ 0\ 0\ 0\ 1\ 1\ 0\ 1 \\
\underline{1\ 0\ 1\ 0\ 1\ 0\ 0\ 1}
\end{array}
$$

Sum   1 0 0 1 1 0 1 1 0

Carry   1 0 0 0 1 0 0 1

6.5 Negative numbers

There are three common ways of representing negative numbers in binary.

6.5.1 Sign plus value

This is similar to writing a decimal number with a + or − sign; +84 or −356 for example. A sign plus value representation usually employs an additional bit to indicate the sign, which is 0 for positive numbers, and 1 for negative numbers.

    0 0 1 0 1   represents decimal +5
    0 1 1 0 0   represents decimal +12
    1 0 1 0 1   represents decimal −5
    1 1 1 0 0   represents decimal −12

Sign plus value is obviously simple, but is not particularly convenient when arithmetical operations have to be performed.

6.5.2 Ones complement

Ones complement again uses the top bit to indicate the sign, being 0 for positive numbers and 1 for negative numbers. For negative numbers, however, the value part of the number is complemented:

0 0 1 0 1 represents decimal +5
0 1 1 0 0 represents decimal +12
1 1 0 1 0 represents decimal −5
1 0 0 1 1 represents decimal −12

Zero causes a problem in ones complement, as it can be represented in two ways. With four-bit numbers, for example, 0000 is 'positive' zero and 1111 is 'negative' zero. This anomaly is avoided by the next representation, called twos complement.

6.5.3 Twos complement

Twos complement is formed by simply adding 1 to the ones complement negative number.

(a)   0 0 0 1 1   +3
       1 1 1 0 0   −3 in ones complement
       1 1 1 0 1   −3 in twos complement

(b)   0 1 0 1 0   +10
       1 0 1 0 1   −10 in ones complement
       1 0 1 1 0   −10 in twos complement

(c)   0 1 0 1 0 1 1 +43
       1 0 1 0 1 0 0 −43 in ones complement
       1 0 1 0 1 0 1 −43 in twos complement

There are two distinct advantages in using a twos complement representation. Firstly there is one distinct value for zero; forming the ones complement of zero and adding one simply gives zero again. Secondly, and more importantly, addition of a positive number and a twos complement negative number of the same magnitude gives zero; for example:

       0 1 0 1 0 1 1   +43
       1 0 1 0 1 0 1   −43 twos complement
       X 0 0 0 0 0 0 0   top bit is lost giving zero

This means that a twos complement representation allows subtraction to be done by addition of a negative number. For example:

```
  0 1 1 0 0      +12
  1 1 1 0 1      −3 in twos complement
X 0 1 0 0 1      top bit is lost giving +9 as the correct answer
```

The equivalent decimal sum is:

$$12 - 3 = 12 + (-3) = 9$$

Subtraction is dealt with in some detail in section 6.6.

6.5.4 General comments

Three ways of representing negative numbers have been described; sign magnitude, ones complement and twos complement. It should be noted that positive numbers are identical in all three systems; it is only the representation of negative numbers (and zero) that differ.

Sign magnitude is the simplest to use where numerical values have to be input or displayed. It is also useful where digital to analog converters are used (DACs, see chapter 12). Arithmetic is difficult.

Twos complement is mandatory when extensive arithmetic is required. It is by far the commonest representation. Conversion to and from positive numbers is somewhat involved, however, requiring some arithmetic.

Ones complement is often found to be a useful compromise. Conversion to and from positive numbers is straightforward, and conversion between one and twos complement is relatively easy. Some computers store numbers in ones complement form and convert to twos complement for arithmetic operations.

Care should be taken with systems using signed binary numbers, because it is not always obvious what a number is representing. An eight-bit number, for example, can represent from 0 to 255 if it is taken to be unsigned, or −128 to + 127 if it is taken to be signed twos complement. The number 1101 0011 can be taken as unsigned 211 or signed −45. Which is correct will depend on the context.

6.6 Subtraction

Subtraction is nearly always performed by taking the twos complement of the subtrahend and performing a normal binary addition. It is useful to consider the similar decimal procedures. For example:

$$
\begin{array}{r}
42 \\
+\ 76 \\
\hline
\cancel{1}18
\end{array}
\quad \text{top digit lost gives answer 18}
$$

76 has behaved like −24, and is said to be the nines complement of 24. To form the nines complement of a number, it is subtracted from one hundred. If we wish to perform the subtraction 24 − 15 the steps would be:

Nines complement of 15 is 85

$$
\begin{array}{r}
24 \\
+\ 85 \\
\hline
\cancel{1}09
\end{array}
$$

Dropping the top digit gives the correct answer 9.

Binary subtraction uses the same principle. The same sum in binary is 11000 (that's 24) minus 1111 (that's 15):

Twos complement of 15 is 110001
Perform the addition

$$
\begin{array}{ll}
\begin{array}{r}
0\ 1\ 1\ 0\ 0\ 0 \\
1\ 1\ 0\ 0\ 0\ 1 \\
\hline
\cancel{1}0\ 0\ 1\ 0\ 0\ 1
\end{array}
&
\begin{array}{l}
+24 \\
-15 \text{ in twos complement}
\end{array}
\end{array}
$$

dropping the top digit gives 001001, which is 9.

Another example: 42 − 24

42 in binary is 101010
24 in binary is 011000

In twos complement we get

$$
\begin{array}{ll}
\begin{array}{r}
0\ 1\ 0\ 1\ 0\ 1\ 0 \\
1\ 1\ 0\ 1\ 0\ 0\ 0 \\
\hline
\cancel{1}0\ 0\ 1\ 0\ 0\ 1\ 0
\end{array}
&
\begin{array}{l}
+42 \\
-24
\end{array}
\end{array}
$$

the answer is 10010 which is the correct answer of 18.

Using twos complement, the correct answer is obtained even when the result is negative. For example: 8 − 11

8 in binary is 1000
11 in binary is 1011

In twos complement

```
0 1 0 0 0          +8
1 0 1 0 1          −11
1 1 1 0 1
```

The most significant bit is 1 so the result is negative. To get back to the value we subtract 1 (this gets back to ones complement):

```
1 1 1 0 0
```

Complementing this gives the value:

```
0 0 0 1 1  which is 3
```

1 1 1 0 1 is therefore −3, which is the correct answer for 8 − 11.

When twos complement arithmetic is being used, care must be taken to avoid incorrect answers caused by arithmetic overflow. Suppose we are using five-bit twos complement and we wish to add 14 and 13. In binary 14 is 01110; 13 in binary is 01101. Adding these we get:

```
0 1 1 1 0          +14
0 1 1 0 1          +13
1 1 0 1 1
```

The most significant bit indicates that the result is negative (it is −5 actually). In five-bit unsigned binary 11011 is, of course, the correct answer of 27, but in five-bit twos complement it is the incorrect answer of −5. When twos complement arithmetic is being used the number of bits should be chosen to ensure overflow cannot occur, or overflow detection circuits should be included.

In the above example, six-bit twos complement would give the correct answer:

```
0 0 1 1 1 0          +14
0 0 1 1 0 1          +13
0 1 1 0 1 1          Result 27, correct
```

Overflow errors can also occur when two negative numbers are added and give an apparently positive result.

Subtraction can be performed in ones complement representation by a technique called 'end around carry'. The digit which would normally be lost off the top end of a binary subtraction is added back in at the bottom (least significant) end to give the correct result. For example, we will evaluate $8 - 5$ in ones complement. 8 in binary is 01000. Ones complement -5 is 11010. The subtraction is:

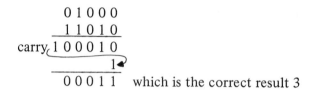

```
        0 1 0 0 0
        1 1 0 1 0
carry 1 0 0 0 1 0
                1
        ─────────
        0 0 0 1 1    which is the correct result 3
```

Another example, this time giving a negative result; $9 - 11$:

```
        0 1 0 0 1    +9
        1 0 1 0 0    −11 in ones complement
carry 0 1 1 1 0 1
                0
        ─────────
        1 1 1 0 1    which is −2 in ones complement
```

Subtraction in sign/magnitude representation is not easy to perform directly. When subtraction is required, a sign/magnitude number is normally converted to one of the other representations first.

6.7 Multiplication and division

Most binary arithmetic involves just addition and subtraction. The operations of multiplication and division are relatively rare. Although the theory and the resulting circuits are straightforward and easy to understand, multiplication and division both require a considerable number of ICs, even with LSI. It is interesting to note that most micro and mini computers do not include multiplication and division in their machine code instructions (although both can, of course, be performed by subroutines). LSI devices are described in section 8.13.

Binary multiplication is analogous to decimal multiplication. A typical decimal multiplication could be:

```
    456  multiplicand
    123  multiplier
  ─────
  45600
   9120
   1368
  ─────
  56088
```

The multiplicand is multiplied by each digit of the multiplier in turn, and the part results added.

In binary multiplication there are only three possible digit combinations:

$0 \times 0 = 0$
$0 \times 1 = 0$
$1 \times 1 = 1$

A typical binary multiplication could be:

```
      1 0 1 1   Multiplicand decimal 11
      1 0 0 1   Multiplier decimal 9
1 0 1 1 0 0 0   Comment 1011 X 1 = 1011
  0 0 0 0 0 0   Comment 1011 X 0 = 0000
    0 0 0 0 0
      1 0 1 1   Comment 1011 X 1 = 1011
1 1 0 0 0 1 1   Result decimal 99
```

Note that multiplication of two four-digit binary numbers has given a seven-digit result. In fact, multiplication of two four-digit numbers can give an eight-digit result as the example below illustrates:

```
      1 1 0 1   Decimal 13
      1 1 0 0   Decimal 12
1 1 0 1 0 0 0
  1 1 0 1 0 0
    0 0 0 0 0
      0 0 0 0
1 0 0 1 1 1 0 0   Decimal 156
```

In a similar way, multiplication of two two-digit decimal numbers can give a four-digit result.

Multiplication of signed numbers causes problems, because the sign of the result must also be computed according to the usual rules:

$(+) \times (+) = (+)$; $(+) \times (-) = (-)$; $(-) \times (-) = (+)$

Usually, in twos complement multiplication, both numbers are converted to unsigned binary, and multiplied. The sign computation is then done to evaluate the sign of the result. If the sign is negative, the result is converted to twos complement.

Division of binary numbers is very rare, although it is performed in the same way as decimal division. For example $25/5 = 5$ in binary is:

```
            1 0 1
1 0 1  )1 1 0 0 1
        1 0 1
          1 0 1
          1 0 1
          0 0 0
```

Division by multiples of 2 (i.e. 2, 4, 8, 16, etc.) is easily achieved by removing the least significant digit for each power of two. For example:

1 1 0 0 0	24
1 1 0 0	12
1 1 0	6
1 1	3

A circuit called a shift register described in section 7.6, can be used to perform division (and multiplication) by powers of two.

Division is sometimes performed by the iterative process of repeated subtraction. The divisor is subtracted from the dividend until the result is zero or negative. The number of subtractions performed is then the result. We can demonstrate the technique by dividing 12 by 4 in decimal:

```
    12
 -  4 =   8
       -  4 =   4
            -  4 = 0   3 subtractions; 12/4 = 3
```

The circuitry to perform repeated subtractions is simple, but slow.

6.8 Binary fractions

Most binary numbers are integers (whole numbers). Very rarely, binary fractions may be encountered. Binary fractional digits represent negative powers of two:

$$2^0 = 1.0 \qquad\qquad 0.111$$
$$2^{-1} = 0.5$$
$$2^{-2} = 0.25$$
$$2^{-3} = 0.125$$
etc.

The binary number 0.101 therefore represents 0.625 decimal.

A fractional decimal number is converted to binary by repeated multiplication by two and recording the carries. For example, 0.6875:

$$0.6875 \times 2 = 1.375 = 0.375 + \text{carry } 1$$
$$0.375 \quad \times 2 = 0.75 \ = 0.75 \ + \text{carry } 0$$
$$0.75 \quad \times 2 = 1.50 \ = 0.5 \quad + \text{carry } 1$$
$$0.5 \quad \times 2 = 1.0 \ \ = 0 \quad\ + \text{carry } 1$$

$$0.6875 = 0.\ 1\ 0\ 1\ 1$$

Usually a fractional decimal number will not convert precisely to a fractional binary number, and the conversion is terminated after a predetermined number of places.

6.9 Other codings

6.9.1 Introduction
Binary coding is the most efficient coding for logic circuits. Other codings have special advantages in some circumstances. The commonest of these alternative codings are binary coded decimal (BCD), cyclical (Gray) codes and XS3 code.

6.9.2 Binary coded decimal (BCD)
A single decimal digit can take any value between 0 and 9. Four binary digits are therefore required to represent a single decimal digit. In BCD, a decimal number is represented by four bits for each decimal digit. For example:

7	4	0	6	9
0111	0100	0000	0110	1001

The BCD representation of 74069 is therefore
111 0100 0000 0110 1001.

BCD is not as efficient as pure binary. Four bits in binary can represent 0 to 15. Four bits in BCD only represent 0 to 9 (1010, 1011, 1100, 1101, 1110, 1111 representing 10 to 15 are not used). A BCD number is always longer than its binary equivalent. For example, 99 in BCD is 1001 1001 which needs 8 bits. In straight binary 99 is 110 0011 which needs seven bits. For larger numbers the difference is more marked.

BCD is very useful where decimal numbers are to be read from, say, decade switches, or numbers are to be displayed on numerical indicators. With BCD each decimal digit is independent. Conversion from decimal to BCD and BCD to decimal is simple because only the binary representations for 0 to 9 need to be considered.

Arithmetic in BCD causes complications because two cases have to be considered. Firstly, if the sum of two digits comes to 9 or less, straight binary addition and BCD addition gives the same result. For example

	Binary	BCD
4	0100	0100
+ 3	0011	0011
7	0111	0111

If the sum of two digits comes to more than 9, BCD must generate a carry. The BCD and binary addition do not give the same result. For example:

	Binary	BCD
5	0101	0101
+ 7	0111	0111
12	1100	00010010

The rules for adding two BCD digits are:

(a) Add digits using normal binary arithmetic.
(b) If result is greater than nine, add six to the result (which gives the correct result and generates the carry).

To demonstrate the technique, we will use BCD to perform the addition:

```
 458
 716
1174
```

In BCD the numbers to be added are:

```
458    0100  0101  1000
716    0111  0001  0110
```

These are added using straight binary arithmetic.

```
0100  0101  1000
0111  0001  0110
1011  0110  1110
```

The least significant digit is decimal 14, which is greater than 9, so we add 6:

```
 1110          14
  110           6
10100
```

We have generated a carry to the second digit which becomes 0111. This is 7, which is less than 9, so no correction is needed.

The most significant digit is 1011. This is decimal 11, so we add 6:

```
 1011
  110
10001
```

The addition generated a carry into the thousands digit, so the BCD result is:

```
0001  0001  0111  0100
   1     1     7     4
```

which tallies with the decimal result.

BCD arithmetic is relatively rare. It is usually more convenient to convert inputs to binary, use binary for any computation, and revert to decimal for outputs.

6.9.3 2421 code and XS3 code
The BCD code described above is known as 8421 code because the binary digits are weighted 8, 4, 2 and 1. Other BCD codes are possible by using other weightings or codings.

Two such codes are the 2421 code (so called because the digits are weighted 2, 4, 2 and 1) and the XS3 code (for excess 3; it is BCD code plus 3). These are compared with 8421 BCD below:

Decimal	BCD	2421	XS3
0	0000	0000	0011
1	0001	0001	0100
2	0010	0010	0101
3	0011	0011	0110
4	0100	0100	0111
5	0101	0101	1000
6	0110	0110	1001
7	0111	0111	1010
8	1001	1110	1011
9	1010	1111	1100

2421 and XS3 code have some advantages in specific circumstances (a 2421 counter is easy to build, and some arithmetic is simpler in XS3) but generally BCD can be taken to be 8421 code unless otherwise stated.

6.9.4 Unit distance (cyclic) codes

Figure 6.3 shows a possible application of binary coding. The position of a shaft is to be measured to 1 part in 16 by an optical grating moving in front of four photo electric cells (PECs). The circuit outputs give a binary representation of the shaft position.

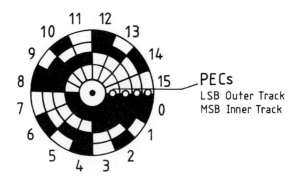

Fig. 6.3 Shaft encoder with binary coding.

Consider what may happen as the shaft moves from position 7 (0111) to position 8 (1000). It is unlikely that all bits would change together so we could get:

0111 to 0000 to 1000
or　　　0111 to 1111 to 1000

or any other combination of four bits. These possible incorrect intermediate states can be avoided by using a code in which only one bit changes between adjacent positions. Such a code is called a unit distance code or cyclic code.

The commonest code is the Gray code which is shown in four-bit form below:

Decimal	Gray
0	0000
1	0001
2	0011
3	0010
4	0110
5	0111
6	0101
7	0100
8	1100
9	1101
10	1111
11	1110
12	1010
13	1011
14	1001
15	1000

base 6 cyclic　　　decimal cyclic　　　symmetrical

It will be noted that the code is reflected about the centre. A unit distance code to any even base can be obtained by taking an equal number of combinations above and below the centre. A cyclic code of 6 and 10 are shown above and are redrawn below. The decimal version is known as XS3 cyclic BCD code.

Cyclic code of 6

0	0111
1	0101
2	0100
3	1100
4	1101
5	1111

XS3 cyclic BCD code

0	0010
1	0110
2	0111
3	0101
4	0100
5	1100
6	1101
7	1111
8	1110
9	1010

Gray codes are sometimes called reflective codes because of this symmetrical property.

Gray codes can be constructed to any length, and to any even base. Codes are built up by using the symmetry property. For example, a five-bit code would have symmetry at the 15/16 point as below:

	etc.	
	13	01011
	14	01001
	15	01000
symmetry	16	11000
	17	11001
	18	11011
	etc.	

Conversion between Gray codes and binary/BCD is described in section 8.11.

CHAPTER 7

Counters and Shift Registers

7.1 Introduction

This chapter examines devices used for counting. Comparison with electromagnetic counters shows that there are basically two counting applications.

The first, and obvious, use is the counting, or totalising, of external events which are usually presented as pulses. Frequency meters, the ramp ADC described in section 12.5.3 and traffic recorders are typical counting applications.

The second, and not so obvious, use of a counter is to divide the frequency of a pulse train to give a new lower frequency. The new frequency is some integer division of the original frequency. A typical application is the VDU timing chain described in section 9.7, where various frequencies down to 50 Hz are produced from a single oscillator operating at over 1 MHz.

Closely related to counters are devices known as shift registers. These are described in section 7.6.

7.2 The toggle flip-flop

The toggle, or T, flip-flop was described briefly in section 4.5.7, where it was shown that a JK flip-flop connected as in fig. 7.1a or a D-type flip-flop connected as in fig. 7.1b will toggle on clock pulse inputs as shown in fig. 7.1c. Dependent on the type of JK or D-type used, toggling can occur on the positive or negative edge of the input. Usually negative edge triggering is used in counting applications.

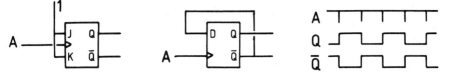

Fig. 7.1 The toggle flip-flop. (a) JK based. (b) D-type based. (c) Waveforms.

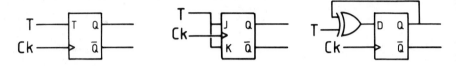

Fig. 7.2 The T-type toggle flip-flop. (a) Symbol. (b) JK version. (c) D-type version.

An alternative form of toggle flip-flop has a T input as well as the clock input. If the T input is 1, the outputs will toggle on clock pulses. If T = 0 the clock pulses have no effect. This form of the toggle flip-flop can be achieved directly with the JK flip-flop as in fig. 7.2b or a D-type with additional gating as in fig. 7.2c.

7.3 Ripple counters

7.3.1 A binary up counter
A binary counter which can only count up is conceptually the simplest counting circuit. A three-bit counter will be described, although the circuit development can be extended to any length of counter.

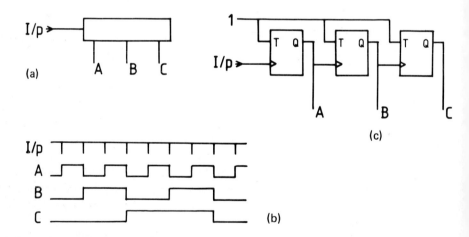

Fig. 7.3 A simple three-bit counter. (a) Counter representation. (b) Waveforms. (c) Implementation with T-types.

A three-bit counter can be represented by fig. 7.3a. Clock pulses are applied to the input and are counted on the three binary outputs ABC. There are certain problems in drawing logic diagrams with counters. Conventionally, signals flow from left to right. Equally conventionally, numbers are drawn with the least significant digits to the right. Eight hundred and fifty three is 853 not 358. Unfortunately, these two conventions conflict in drawing counter circuits. It is usual, but by no means universal, to stay with the convention that signals flow from left to right. This convention has been followed in counter circuits described below. The letters A, B, C, etc. are used for counter outputs, with A being the least significant. Often the terminology Q_a, Q_b, Q_c, etc. is used.

The counter in fig. 7.3a will therefore count on successive clock pulses as below:

Decimal	C	B	A
0	0	0	0
1	0	0	1
2	0	1	0
3	0	1	1
4	1	0	0
5	1	0	1
6	1	1	0
7	1	1	1

A clock pulse when the counter has a count of 7 (binary 111) will take it back to 0 in the same way that a car milometer can go over the top. The counting sequence is also shown on the timing diagram of fig. 7.3b.

Examination of the table above and fig. 7.3b shows that any bit changes state (i.e. toggles) when the preceding less significant bit goes from a 1 to a 0. A simple binary counter can therefore be constructed from three toggle flip-flops as in fig. 7.3c. Each Q output is connected to the clock input of the succeeding stage, so as an output goes from a 1 to a 0, the succeeding stage will toggle. The circuit behaves as a simple three-bit binary counter.

7.3.2 Glitches
Each of the toggle flip-flops in fig. 7.3 has an inherent propagation delay. Consider what happens when we go from 3 (011) to 4 (100):

(i)   Negative edge of clock pulse arrives. Counter state 011;
(ii) A output changes from 1 to 0. Counter state transitorily 010;
(iii) A output changing from 1 to 0 causes B to toggle from 1 to 0. Counter state transitorily 000;

(iv) B output changing from 1 to 0 causes C to toggle from 0 to 1. Counter reaches final state of 100 (4).

Each step occupies a time equivalent to the propagation delay of one toggle flip-flop, and is summarised in fig. 7.4. Going from a count of 3 to a count of 4, transient states of 2 and 0 have appeared. Because the effect of a clock pulse 'ripples' through the counter from the least significant end, the circuit of fig. 7.3c is often called a ripple counter.

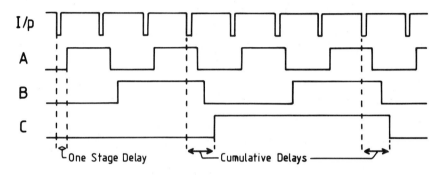

Fig. 7.4 The effect of propagation delays.

The number of transient states and the time taken for the counter to settle to its new value is directly proportional to the number of bits in the counter. The relevance of the transient states depends on the application. If the counter outputs are fed to a decoder, invalid glitches may appear on the decoder outputs. If the decoder output is driving indicator lamps the glitches would not be noticed. If the decoder outputs were selecting sections of a semiconductor memory in a microcomputer, problems would probably ensue.

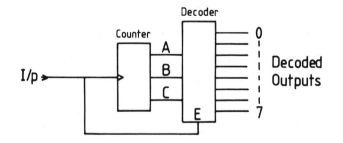

Fig. 7.5 Use of decoder enable input to prevent glitches.

One possible solution would be to gate the decoder outputs with the clock pulse as in fig. 7.5. Decoder outputs are then inhibited during the (short) time that the ripple is progressing through the counter, and the outputs are glitch free. Another solution is the synchronous counter described later in section 7.4.

Logic families with rise times significantly longer than their propagation delay (such as CMOS and most industrial logic) have an in-built immunity to glitches. Problems with glitches are more common in TTL and ECL.

7.3.3 A binary down counter

A down counter is frequently required. In packaging, for example, a required batch quantity may be loaded into a counter which is decremented for each item. When the counter reaches zero the batch is complete.

A three-bit binary counter would behave as follows:

Decimal	C	B	A
7	1	1	1
6	1	1	0
5	1	0	1
4	1	0	0
3	0	1	1
2	0	1	0
1	0	0	1
0	0	0	0

If the counter receives an input pulse when the counter is in the zero state it will go back to the seven state.

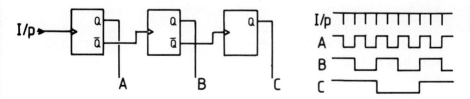

Fig. 7.6 A three-bit down counter. (a) Circuit diagram. (b) Timing waveforms.

Examination of the above table shows that any bit changes when the preceding less significant bit changes from a 0 to a 1. A down ripple counter can therefore be constructed by linking the $\bar{Q}$ outputs from each stage to the clock input of the succeeding stage, and taking the counter outputs (ABC) from each Q output as shown in fig. 7.6a. The timing waveforms are shown in fig. 7.6b.

The down ripple counter operates in a similar manner to the up counter and can also suffer from transient false states. Going from 4 to 3, for example, the counter will transiently go $4 - 5 - 7 - 3$. These transient states can be overcome as described in section 7.3.2.

7.3.4 Limitations to speed

There are two constraints on the maximum operating speed at which a ripple counter can operate. The first is the maximum operating frequency of the individual toggle flip-flops. This is determined by the various factors given in sections 1.5.2 and 4.7, and is not dependent on counter length.

The second constraint is the cumulative propagation delays which give the ripple-through effect. If the cumulative delays exceed the clock period, the counter state can never be read sensibly. The cumulative delay is directly proportional to counter length, so it is quite feasible to have a counter which operates incorrectly whilst still within the manufacturer's 'maximum operating frequency'.

7.3.5 Frequency division

Examination of fig. 7.3 shows that the output frequency of the most significant bit (C) is precisely one-eighth of the input frequency. A simple ripple counter will act as a frequency divider.

If we define the division ratio as

$$N = \frac{\text{fin}}{\text{fout}}$$

then for m binary stages $N = 2^m$. For a single toggle flip-flop, $m = 1$ and $N = 2$. For three stages, $m = 3$ and $N = 8$, as we saw above.

A ripple counter operating as a frequency divider is limited solely by the maximum operating frequency of the first stage. The output will, however, be time shifted by the total ripple propagation delay. If this delay is unacceptable, a synchronous counter should be used.

The output of a binary counter has a unity mark/space ratio regardless of the input frequency mark/space ratio providing the input frequency is constant.

7.3.6 MSI ripple counter

Although ripple counters can be constructed from D-type or JK flip-flops, it is often more convenient to use MSI counters. Typical of these is the TTL 7493 four-bit counter and the CMOS 4024 seven-bit counter.

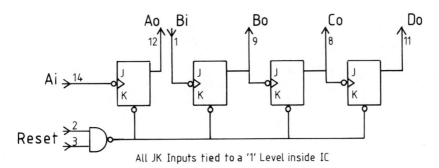

All JK Inputs tied to a '1' Level inside IC

Fig. 7.7 7493 four-bit binary ripple counter.

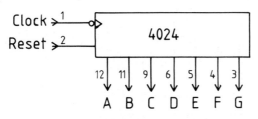

Fig. 7.8 4024 seven-bit binary ripple counter.

The 7493, shown in fig. 7.7, consists of a three-bit counter and an independent one-bit counter. To give a four-bit counter, output Ao is connected to input B. The device operates as an up counter. The count is reset to zero by taking both reset pins to a 1, a facility that can be used to provide a counter to any base, as we shall see in the following section.

The LS version is guaranteed to operate at 32 MHz into the A input and 16 MHz into the B input. Total ripple delay from A input to D output is 70 ns maximum. This implies a counting limit (as opposed to a frequency division limit) of about 15 MHz.

The 4024, shown in fig. 7.8, is a straight seven-bit counter with a single reset input (input 1 to reset). The actual counter is connected as a down counter, but inverter/buffers on the output make the actual counter outputs go *up*. (The actual internal connections are a bit more complex than fig. 7.8 implies, but this is of no concern to the user. Fig. 7.8 defines the way the device behaves.)

The maximum operating frequency and ripple delay depend (like all CMOS devices) on the supply. At 5 V, f max is 5 MHz, and the delay 100 ns per stage (200 ns for the first stage). The device can just operate at 1 MHz. At 15 volts, f max is 18 MHz and the propagation delay per stage falls to 50 ns for all stages.

7.3.7 Counting to other bases

Counting to bases other than two is often required. With ripple counters, any counting base can be achieved with variations of the technique shown in fig. 7.9a. The counter output is decoded by some external logic. When the counter reaches the desired maximum count the decoder output resets the counter.

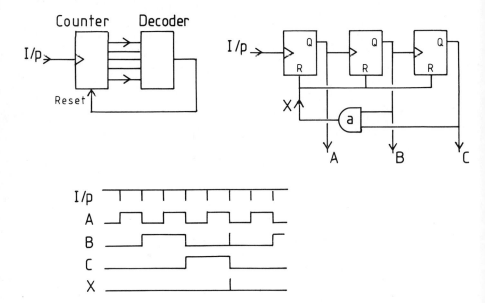

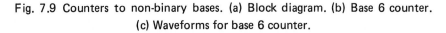

Fig. 7.9 Counters to non-binary bases. (a) Block diagram. (b) Base 6 counter. (c) Waveforms for base 6 counter.

A base 6 counter using the above principle is shown in fig. 7.9b. A three-bit binary counter will count to 8, so it will suffice to count to 6. A base 6 counter will be required to count 0, 1, 2, 3, 4, 5, 0, 1, etc. The three-bit counter is therefore required to reset on a count of six. Gate a gives a 1 out when counter outputs B and C are both 1, i.e. when a count of 6 is reached. The gate output resets the counter to zero. The timing waveforms are shown in fig. 7.9c.

Note that a transient count of 6 is given, with a duration of approximately the propagation delay of one stage. If this is likely to cause a problem, the decoded counter outputs could be disabled as shown on fig. 7.5 before use elsewhere.

It may be thought that a possible 'race' condition exists, in that gate a is resetting the counter and removing signals B and C which provide gate a with its inputs. In practice, gate and flip-flop propagation delays ensure that the pulse out of gate a is adequate to perform the reset action. If doubts exist, the pulse width can be stretched by a capacitor, as in fig. 7.10a, or latched by an external RS flip-flop, as in fig. 7.10b. In the latter circuit (which is totally foolproof) the RS flip-flop is set by the decoding gate, and reset when the count input goes positive again.

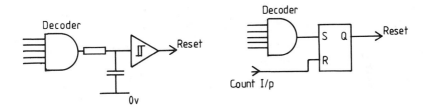

Fig. 7.10 Prevention of race conditions. (a) Pulse stretcher. (b) Pulse latch.

Many binary ripple counter ICs incorporate gating on the reset inputs. The 7493, described in section 7.3.6 has a two-input AND gate on its reset inputs which can be used directly to implement gate a on fig. 7.9 with no external logic.

Set inputs can also be used to force other count patterns. The 2421 code was described in section 6.9.3:

Decimal	D	C	B	A
0	0	0	0	0
1	0	0	0	1
2	0	0	1	0
3	0	0	1	1
4	0	1	0	0
5	0	1	0	1
6	0	1	1	0
7	0	1	1	1
8	1	1	1	0
9	1	1	1	1

This can be implemented by fig. 7.11. When a count of 8 is reached, output D goes to a 1 which in turn forces outputs B and C to a 1 as well. The counter therefore goes from 0111 (7) to 1000 (transiently) to 1110 (8). The set inputs should be edge triggered.

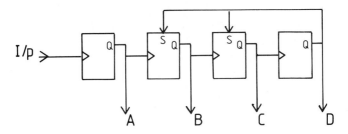

Fig. 7.11 2421 code counter.

Set and reset inputs can be used together. The XS3 code was also introduced in section 6.9.3.

Decimal	XS3
0	0011
1	0100
2	1010
3	1100
4	1110
5	1000
6	1001
7	1010
8	1011
9	1100

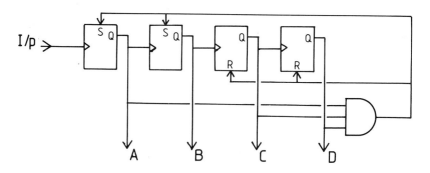

Fig. 7.12 XS3 code counter.

This behaves as a normal counter except when it goes from 9 (1100) back to 0 (0011). On fig. 7.12, the counter is allowed to count to ten (1101) which is decoded by gate a, whose output resets outputs C, D and sets outputs A and B. The counter therefore counts 1100

(9), 1101 (transiently), 0011 (zero).

Counters based on fig. 7.9 do not usually have a unity mark/space ratio on the most significant output, which could preclude their use in some frequency division applications. If a unity mark/space ratio is required, a counter similar to fig. 7.9 should be followed by a divide-by-two toggle flip-flop. A unity mark/space decade counter, for example, could be obtained by a three-bit divide by five counter followed by a divide-by-two toggle flip-flop. Similarly, a unity mark/space divide-by-six counter could be constructed by a divide-by-three counter followed by a toggle flip-flop.

7.3.8 Up/down counter

At first sight, constructing a ripple counter with up/down selection might be thought a relatively simple task. In the up mode, each clock input is connected to the Q output of the preceding stage. In the down mode, each clock input is connected to the $\overline{Q}$ output of the preceding stage.

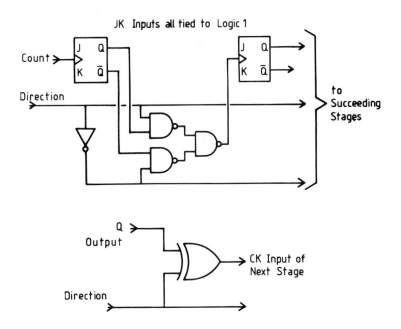

Fig. 7.13 Up/down ripple counter. (a) Basic principle. (b) XOR stage linking.

A tentative design for an up/down ripple counter could be similar to that in fig. 7.13. An AND/OR combination between stages selects Q or $\overline{Q}$ outputs according to the state of the direction selection input. If this input is a 1, the Q outputs are selected and the counter counts up; if it is 0, the counter counts down. (The changeover could be done with fewer gates by using an XOR gate as in fig. 7.13b, but fig. 7.13a better illustrates the principle.)

Although fig. 7.13 will work after a fashion, it has a major short-coming. When the direction is changed, some clock inputs will go from a 1 to a 0, and the corresponding output will change. Each direction change will be accompanied by an unpredictable change in counter output.

One (somewhat inelegant) solution is to slug the direction line with a capacitor, so that edge speeds on direction changes are too slow to toggle the flip-flops. This approach does work, but makes the counter rather noise-prone.

Although it is possible to design a foolproof up/down ripple counter, it is usually simpler, and cheaper, to use the synchronous circuits described below.

7.4 Synchronous counters

7.4.1 Introduction

The simplest synchronous binary counters are based on the toggle flip-flop of fig. 7.2 described in section 7.2. The T input controls the toggle input. If T is 1, the flip-flop will toggle on application of the clock pulse. If T is 0, the flip-flop is unaffected by the clock pulse.

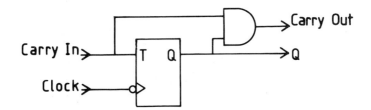

Fig. 7.14 Basic circuit for synchronous counter.

In fig. 7.14, the T flip-flop is used with an AND gate to produce a single-bit counter. This has two inputs, carry-in and clock, and two outputs, carry-out and the Q output. The carry out is 1 when the carry in is 1 and the Q output is 1. The flip-flop will toggle when the carry in is 1.

To construct a counter, several single-bit counters are cascaded as in fig. 7.15. Note that the clock input is common to each stage, and the carry out from each stage is connected to the carry in of the succeeding stage. Each T input will be 1 when *all* the least significant stages are 1, which is the condition required for a counter stage to toggle. The least significant carry in is connected to a 1 level, because the least significant stage toggles on each input pulse.

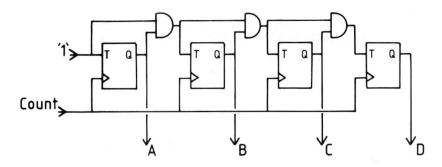

Fig. 7.15 Series connected synchronous counter.

The counter is truly synchronous; all toggling stages change together. There *is* a cumulative propagation delay down the carry stages, but this solely determines the maximum operating speed. Provided the constraints on speed outlined below are observed, the counter will operate synchronously to any required number of bits.

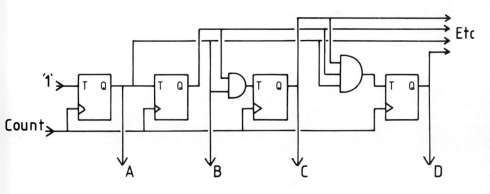

Fig. 7.16 Parallel connected synchronous counter.

There are two constraints on speed. The first is the toggle speed of the actual flip-flops; the second is the propagation delay of the various carry stages. The cumulative delay must be less than the clock period. The counter of fig. 7.15 is known, for obvious reasons, as a series connected counter. A synchronous counter can also be constructed with the so-called parallel connection of fig. 7.16. In this circuit, each stage generates its own carry in signal from the preceding Q outputs. The carry in signals only suffer one gate propagation delay, at the expense of multi-input gates. The circuit of fig. 7.16 can therefore count at significantly higher speeds than the series connected circuit of fig. 7.15, and should be used for very long, or high speed, counters.

Because all toggling stages change together, a synchronous counter output can be decoded without glitches occurring from ripple delays. Glitches may, however, be introduced by the decoder itself. This problem is discussed in section 3.4.

Construction of a synchronous down counter is similar, except carry out signals are generated from $\bar{Q}$ and carry out (for series connections), or $\bar{Q}$ signals gated to produce carry in (for parallel connection). To be strictly accurate, the carry signals should be labelled 'borrow in' and 'borrow out'. Down counters are subject to the same speed constraints as up counters.

7.4.2 Up/down binary counter

A synchronous binary counter with selectable direction is shown in fig. 7.17. This is simply a combination of a synchronous up counter, and a synchronous down counter, both series connected. If the direction line is a 1, gates 1, 2, 3 are enabled, and the up carry outs are passed to the T inputs of the succeeding stage. If the direction line is a 0, gates 4, 5, 6 are enabled, and the borrow outs are passed on. The counter correspondingly counts up when the direction line is 1 and down when the direction line is 0.

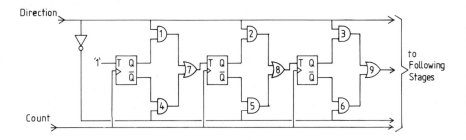

Fig. 7.17 Up/down synchronous counter.

The synchronous counter does not react unpredictably to changes in the direction line. The flip-flops only react to the state of the T inputs at the time of the clock edge. It is, however, good practice not to change the state of the direction line during the time that the clock pulse is high. A randomly occurring direction signal can be synchronised by an external D-type or JK flip-flop as in fig. 7.18.

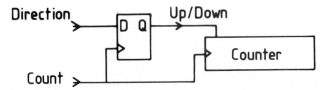

Fig. 7.18 Synchronising direction line.

Figure 7.17 is a series connected reversible counter. It is, of course, feasible to construct a parallel connected counter, and fig. 7.19 shows a typical circuit. By using positive and negative gates ('a' is a conventional AND gate, 'b' is a negative NAND gate) only the Q output of each stage is used. Parallel connected counters can operate at higher speeds than series connected counters for the reasons outlined above.

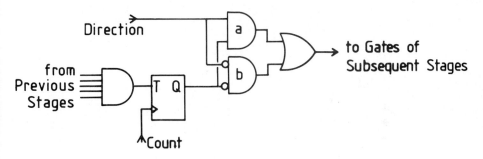

Fig. 7.19 Basics of synchronous parallel connected up/down counter.

7.4.3 State diagrams

In the design of counters, it is often convenient to represent the counter operation by a state diagram. These show all possible states of the counter, and the transitions between the states.

For binary counters the state diagrams are relatively trivial. Figure 7.20 shows a state diagram for a reversible three-bit counter. The state diagram shows the simple transitions through the eight binary states.

State diagrams are particularly useful in visualising the operation of counters which do not use all possible counter states. In fig. 7.12 a circuit for an XS3 counter was shown. This used only ten of the 16 possible states of a four-bit counter. The state diagram of fig. 7.21 shows what happens in all 16 counter conditions, as the counter

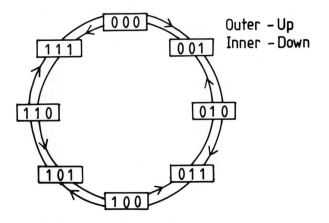

Fig. 7.20 State diagram for three-bit reversible counter.

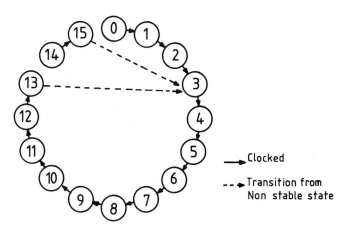

Fig. 7.21 State diagram for XS3 counter.

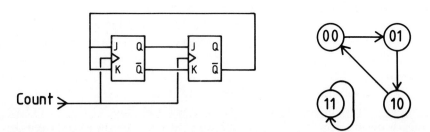

Fig. 7.22 Faulty design for base 3 counter. (a) Circuit diagram. (b) State diagram.

could end up in one of these states at, say, power up.

Figure 7.21 shows that states 1101 and 1111 are not stable; the counter will immediately flip to state 0011. State 1110 will go to state 0011 at the next count pulse. States 0000 to 0010 will count up to 0011 and thereafter will count correctly. Figure 7.21 tells us that in whatever state fig. 7.12 happens to power up, it will eventually count correctly in XS3 code. Such a counter is called a self-starting counter.

Consider the synchronous counter of fig. 7.22a. This is a possible design for a divide-by-three counter. Analysis of the circuit will show that it has the state diagram of fig. 7.22b. As can be seen, the counter will operate correctly if it is in one of the used states. Should it ever attain the unused 11 state, however, it will stay in that state and not resume counting. Such a counter is called a non self-starting counter.

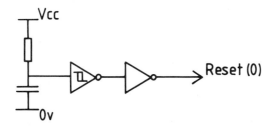

Fig. 7.23 Power up reset.

Although an external reset button can be provided, or an automatic power up reset as in fig. 7.23 incorporated in the circuit, non self-starting counters should be avoided. Automatic power up resets often will not perform reliably with a 'noisy' supply interruption, and manual intervention cannot be relied on. The designer should always use a state diagram to ensure his counter is self-starting.

7.4.4 Synchronous counters to other bases

This section outlines the procedures for designing a parallel connected counter for any required sequence. Usually an MSI counter, such as those described in section 7.4.8, will be more efficient and economical and the designer should always check to see if a standard device is available.

The design procedure below is based on JK flip-flops and is illustrated by the four-bit counter shown in fig. 7.24. The eight counters outputs (Q and $\bar{Q}$ for ABCD) indicate the current state. Eight combinational logic circuits provide the J and K inputs necessary to put the counter in the next state on receipt of the next clock pulse.

The design procedure is:

(a) Draw up a table showing all the required states of the counter and the required next state in the sequence.

(b) From this table, deduce the J and K inputs to be produced at each counter state.

(c) Design the combinational logic required to give these JK inputs.

(d) Draw a full state diagram to check if the counter is self-starting. If not, modify the design.

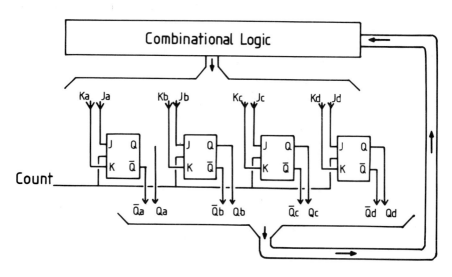

Fig. 7.24 Generalised four-bit synchronous counter.

It is possible at stage (a) to define completely all transitions, including unused states, but this usually leads to an unnecessarily complicated design. Unless the designer is particularly concerned about the route that a counter takes back from unused states, it is usually easier to design an optimal counter and check that it is self-starting. If the design proves to be non self-starting, usually a small obvious change is required.

At stage (b) there are four possible conditions to be considered. If we call the two possible flip-flop states set and reset as below:

Set $\qquad Q = 1, \bar{Q} = 0$
Reset $\qquad Q = 0, \bar{Q} = 1$

we can tabulate the four possible transitions and the JK inputs required to produce them:

Current	Next	J	K	
Set	Set	X	0	J = 0 Unchanged, J = 1 Sets
Set	Reset	X	1	J = 0 Resets, J = 1 Toggles
Reset	Reset	0	X	K = 1 Resets, K = 0 Unchanged
Reset	Set	1	X	K = 0 Sets, K = 1 Toggles

From this table, a table of the required JK inputs can be drawn up. Note the X (don't care) states. These are particularly useful in minimising the combinational logic design at the next stage.

Stage (c) is best achieved by transferring each J and K pattern to a Karnaugh map and producing a minimal design as described in section 3.3.3.

The final stage of checking the design for self-starting simply involves deducing the next state for each of the unused states and drawing the transition diagram. Usually any required corrections will be obvious.

To illustrate this procedure, an 8421 BCD code up counter will be designed.

Stage (a) Tabulate the counter states and the required next states:

Decimal	Present	Next
0	0000	0001
1	0001	0010
2	0010	0011
3	0011	0100
4	0100	0101
5	0101	0110
6	0110	0111
7	0111	1000
8	1000	1001
9	1001	0000

Stage (b) Deduce the necessary JK inputs to give the next state:

Decimal	Present	Next	Jd	Kd	Jc	Kc	Jb	Kb	Ja	Ka
0	0000	0001	0	X	0	X	0	X	1	X
1	0001	0010	0	X	0	X	1	X	X	1
2	0010	0011	0	X	0	X	X	0	1	X
3	0011	0100	0	X	1	X	X	1	X	1
4	0100	0101	0	X	X	0	0	X	1	X
5	0101	0110	0	X	X	0	1	X	X	1
6	0110	0111	0	X	X	0	X	0	1	X
7	0111	1000	1	X	X	1	X	1	X	1
8	1000	1001	X	0	0	X	0	X	1	X
9	1001	0000	X	1	0	X	0	X	X	1

The JK inputs for state 9, for example, are derived as follows:

Stage A   Current Set, Next Reset Ja = X Ka = 1
Stage B   Current Reset, Next Reset Jb = 0 Kb = X
Stage C   Current Reset, Next Reset Jc = 0 Kc = X
Stage D   Current Set, Next Reset Jd = X Kd = 1

Stage (c) Transfer the eight JK required outputs to eight Karnaugh maps (the ABCD inputs are the present counter states). These are shown on fig. 7.25. These are minimised as described in section 3.3.3. Note that stage A gives the simple solution J = K = 1. The corresponding circuit is shown in fig. 7.25b.

Stage (d) The design is checked, both for correct operation and self-starting. This is done by drawing up a table for *all* 16 states.

Current		D	C	B	A	Next		
Decimal	DCBA	JK	JK	JK	JK	DCBA	Decimal	
0	0000	00	00	00	11	0001	1	
1	0001	01	00	11	11	0010	2	
2	0010	00	00	00	11	0011	3	
3	0011	01	11	11	11	0100	4	
4	0100	00	00	00	11	0101	5	Used
5	0101	01	00	11	11	0110	6	
6	0110	00	00	00	11	0111	7	
7	0111	11	11	11	11	1000	8	
8	1000	00	00	00	11	1001	9	
9	1001	01	11	01	11	0000	0	

Current		D	C	B	A	Next	
Decimal	DCBA	JK	JK	JK	JK	DCBA	Decimal
10	1010	00	00	00	11	1011	11
11	1011	01	11	01	11	0100	4
12	1100	00	00	00	11	1101	13	Unused
13	1101	01	00	01	11	0100	4
14	1110	00	00	00	11	1111	15
15	1111	11	11	01	11	0000	0

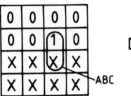

Maps for Ja,Ka all 1

Layout

Fig. 7.25(a)

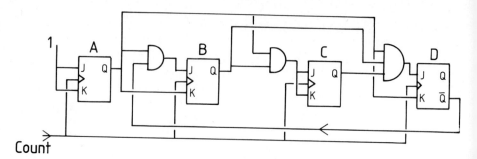

Count

Fig. 7.25 Design of BCD synchronous counter. (a) Karnaugh map. (b) 8421 counter circuit.

From this, the state diagram of fig. 7.26 can be drawn. This shows that the counter operates correctly, and is self-starting in that all unused states eventually enter the correct sequence. If the routes back from unused states are acceptable, the design is complete.

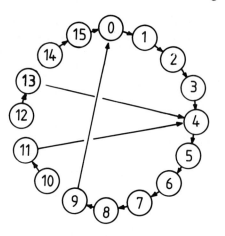

Fig. 7.26 State diagram for 8421 code counter design.

It is usual to find that a design is self-starting, but if not the state diagram will indicate what changes need to be made. Similarly, unacceptable routes can be modified. For example, if the routes from 11, 13 are unacceptable, JcKc logic requires modification.

The above design procedure can be used for any required count sequence, however obscure. It can also be used to design up/down counters which need not have a down sequence the same as the up sequence.

To design an up/down counter, separate up and down logic is designed as described above, and a two-way data selector (controlled by the direction signal) used to determine if the counter takes its JK inputs from the up or the down logic. The resulting design is shown in block diagram form in fig. 7.27.

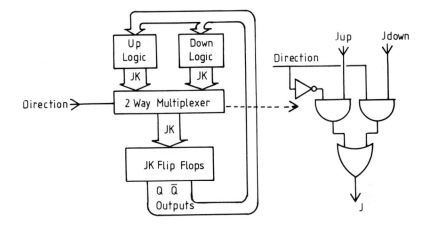

Fig. 7.27 Bidirectional counter design.

7.4.5 Designing with D-types

The design procedure for counters based on D-types is similar to that outlined above for JK flip-flops:

(a) Draw up a table showing all the required states of the counter, and the required next state in the sequence.
(b) From this table, deduce the D inputs required at each counter state. There are, again, four possible conditions,

Current	Next	Required D
Set	Set	1
Set	Reset	0
Reset	Reset	0
Reset	Set	1

Note that there is only one input to provide per stage, and no don't care states.
(c) Design the combinational logic to give these D inputs (using Karnaugh maps).
(d) Draw a full state diagram, and check for self-starting.

A D-type design is shown in block diagram form in fig. 7.28.

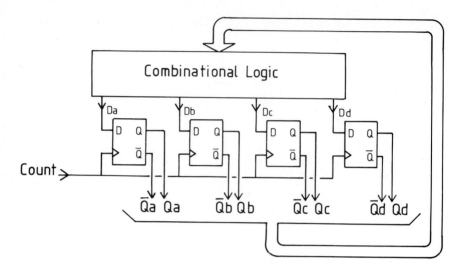

Fig. 7.28 D-type counter design.

7.4.6 A Gray code counter

If the required sequence for a counter differs notably from the straight binary count, the logic designed by the procedures in sections 7.4.4 and 7.4.5 will be correct, but complex. Often the design can be simplified by the use of XOR gates.

A two-bit Gray code counter, with the sequence $00 - 01 - 11 - 10 - 00$, can, for example, be implemented by the circuit of fig. 7.29. The corresponding design with AND/OR gate logic would be far more complex.

7.4.7 Presettable counters

If JK or D-types with direct preset or clears are used, a counter (ripple or synchronous) can be pre-loaded with any required number. The counter in fig. 7.30, for example, has been designed to be forced

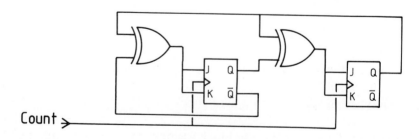

Fig. 7.29 Gray code counter.

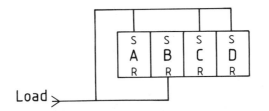

Fig. 7.30 Preloading a counter to 13.

to a count of 13 when the preset line is taken to a 1.

Presettable counters are useful in applications such as batch counting, where a down counter is loaded with the batch size. The counter counts down, and when it reaches zero the batch is complete.

Preloading can be direct (sometimes called a jam load), or gated in some fashion (sometimes called synchronous load).

7.4.8 MSI synchronous counters

Most commonly used count sequences can be obtained in MSI form, and it is usually more economic for a designer to use an MSI counter than build one from scratch using D-type or JK flip-flops. A full list of synchronous counters available in MSI form is given in section 7.7.

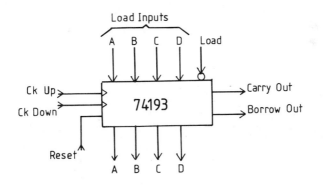

Fig. 7.31 TTL 74193 up/down counter.

A typical device is the 74193 shown diagrammatically in fig. 7.31. This is a four-bit binary up/down counter with individual preset and common clear.

The counter uses a separate clock input for up and down operation rather than a common clock and direction line as described earlier. To count up, a clock pulse is applied to the CKup input. To count down, a clock pulse is applied to the CKdown input. The counter counts on the positive edge, and the unused clock input should be left high.

The carry out and borrow out give a pulse output synchronous with the corresponding clock input when a carry (or borrow) is generated. When counters are cascaded these are connected direct to the clock inputs of the next stage.

Parallel inputs are provided. These are gated via the load signal and the counter outputs assume the state of the ABCD inputs. These inputs overrule the clock. The load input is active low. An active high reset clears all outputs to 0.

The standard and LS version of the 74193 will operate typically at 32 MHz, and the L version at 7 MHz.

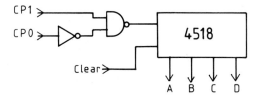

Fig. 7.32 One half of 4518 dual BCD counter.

A typical non-binary counter is the 4518 which comprises two synchronous BCD counters, each identical to fig. 7.32. These are straightforward BCD counters which can be positive edge triggered (on CP0) or negative edge triggered (on CP1). A common clear input resets the counter to zero.

Carry-out decoding is not included, so it is not possible to cascade 4518 counters without external logic.

7.5 Hybrid counters

It is often possible to design an economic counter by mixing synchronous and ripple stages. This technique is often used in MSI so-called ripple counters.

Typical of these is the 7492 divide-by-twelve counter shown in fig. 7.33. This incorporates a synchronous divide-by-three stage and two divide-by-two ripple stages.

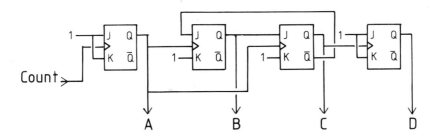

Fig. 7.33 Base 12 hybrid counter (7492).

7.6 Shift registers

7.6.1 Introduction
Shift registers are close relatives of counters, and are found in a wide range of applications. A very simple shift register, constructed from D-type flip-flops, is shown in fig. 7.34a. This circuit could be extended to any desired length.

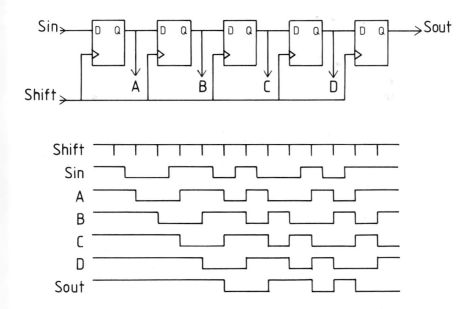

Fig. 7.34 Simple shift register. (a) Circuit diagram. (b) Waveforms.

Data applied to the Serial input, Sin, will move one place to the right on each clock pulse as shown on the timing waveform, fig.

7.34b.

Observation of fig. 7.34b will show that the S output matches Sin, but is delayed by five clock pulses (ABCD are delayed by 1, 2, 3, 4 clock pulses). One use of a shift register is a controlled delay of signals in a synchronous system.

Suppose Sin represents some data in serial form. After four clock pulses, that data will appear in parallel form on ABCD. A shift register can be used to convert a serial signal to a parallel signal. This application is called a serial in, parallel out shift register.

If the four flip-flops are parallel loaded via their preset and clear inputs, and four clock pulses applied, a serial representation of ABCD will appear on Serial out, Sout. A shift register can also be used to convert a parallel signal to a serial signal.

Shift registers are also the basis of many arithmetic circuits. Suppose we have a four-bit shift register, containing:

D	C	B	A
0	1	0	1

This represents decimal five. One place shift to the left gives 1010 which is decimal ten. A one-place shift towards the most significant bit multiplies by two. Conversely, a one-place shift towards the least significant bit divides by two. For example 1100 (decimal 12) when shifted becomes 0110 (decimal six). To avoid confusion shifts towards the MSB will be called a shift up, and shift towards the LSB a shift down.

There are similar problems when drawing shift registers to drawing the counters described in section 7.3. Either conventional flow from left to right or binary positioning must be lost if unnecessarily complex drawing is to be avoided. In this text, left to right logic flow is maintained. As usual, the LSB of a binary or BCD number is denoted by A.

7.6.2 Bidirectional shift register

A bidirectional shift register can be constructed as in fig. 7.35. The operation is fairly straightforward. With the direction line at a 1, each D input is connected to the next output to the left. With the direction line at a 0, each D input is connected to the next output to the right. Clock pulses will then shift data right or left dependent on the state of the direction line.

Parallel data can be loaded directly via the preset and clear inputs. If used in arithmetic circuits the direction line can then be used to select multiplication or division by powers of two.

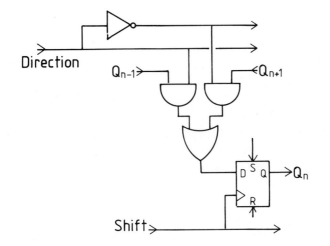

Fig. 7.35 One stage in a bidirectional shift register.

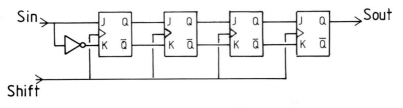

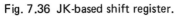

Fig. 7.36 JK-based shift register.

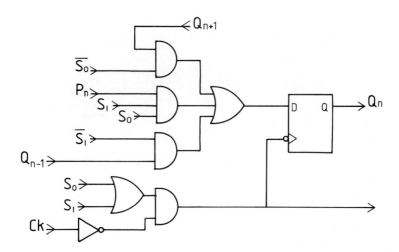

Fig. 7.37 One (of four) stages in the 74194 bidirectional shift register.

7.6.3 JK shift registers

It is also possible to construct shift registers with JK flip-flops as shown in fig. 7.36. Bidirectional shift registers using JK flip-flops require more complex gating, however, so most practical circuits are based on the D-type circuit of fig. 7.35.

7.6.4 MSI shift registers

Most shift registers are, of course, obtained in MSI form. Typical of these is the 74194 shown in fig. 7.37. The 74194 is a four-bit shift register with selectable direction, parallel load and common clear.

The device operation is controlled by the two sense lines, S0, S1 as below:

S1	S0	Mode
0	0	Do nothing
0	1	Shift right
1	0	Shift left
1	1	Parallel load

Note that the parallel load is entered by the clock; it is a synchronous parallel load. Serial data can be entered via the SRin (shift right) or SLin (shift left) according to the direction selected.

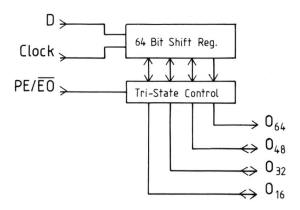

Fig. 7.38 One half of 4517 dual 64-bit shift register. The other half is identical.

MSI shift registers are also available to implement clocked delays. The 4517, for example, provides two 64-bit shift registers, each identical to fig. 7.38. It would be impossible to bring all 64 bits out to IC pins, but bits 16, 32, 48 and 64 are made available. These pins can also be used for data input.

When PE/EO is low, the four O pins are connected to the outputs of stages 16, 32, 48 and 64 and the device behaves as a 64-bit shift register. When PE/EO is high, the 64-bit shift register behaves as four 16-bit shift registers with serial inputs from D, O16, O32, O48; O64 is disabled.

The 4517 can therefore be used to delay a serial signal by 16, 32, 48 or 64 clock pulses. This type of shift register is the basis of some forms of FIFO (first in, first out) memory.

7.6.5 Shift register counters

Shift registers can also be used as counters, but the resulting circuits are not as efficient as straight binary counters. In some applications however (particularly sequencing) they do have certain advantages.

There are two basic types of shift register counter. The first is known as the ring counter. This cycles one bit through the shift register. A four-bit shift register would count:

	D	C	B	A	
0	0	0	0	1	
1	0	0	1	0	
2	0	1	0	0	
3	1	0	0	0	
0	0	0	0	1	etc.

The counter is very inefficient, using four flip-flops compared with the two needed for a binary circuit. The actual circuit for a four-bit ring counter is shown in fig. 7.39. The standard circuit for a ring counter is a NOR gate with inputs from all but the last stage, and the NOR gate output connected to the input of the first stage.

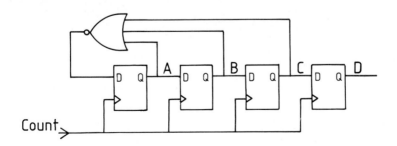

Fig. 7.39 Four-bit ring counter.

The second type of shift register counter is called the Johnson counter (also called a twisted ring or walking ring). The sequence for a three-bit counter is:

	C	B	A
0	0	0	0
1	0	0	1
2	0	1	1
3	1	1	1
4	1	1	0
5	1	0	0
0	0	0	0	etc.

This is more efficient than the ring counter in that six states are produced from three flip-flops. The circuit for a three-bit (divide-by-six) Johnson counter is shown in fig. 7.40.

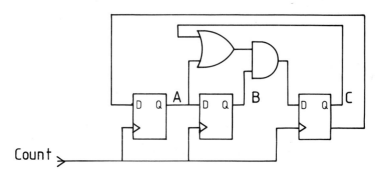

Fig. 7.40 Twisted ring (Johnson) counter.

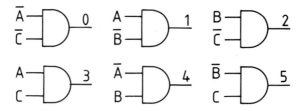

Fig. 7.41 Decoding a Johnson counter.

Johnson counters are widely used for sequencing applications. It can be seen from the table above that only one bit changes at a time (similar in principle to Gray codes). A decoded Johnson counter will be glitch-free. Johnson counters are also easy to decode. All six states can be decoded with two input gates as shown in fig. 7.41.

The CMOS family contains the 4017 which is a very useful divide-by-ten (five-bit) Johnson counter with decoded outputs. (The outputs, actually, will be identical to a 10-bit ring counter output.) The device is shown in block diagram form in fig. 7.42. The device can be triggered on positive edges via CP0, or negative edges via CP1. The carry output can be linked to CP0 on the succeeding stage to make a counter of any desired length.

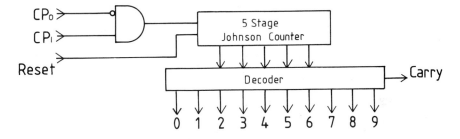

Fig. 7.42 4017 five-stage Johnson counter with decoded output.

7.6.6 Computing shift registers

Computers use shift registers for a variety of logical and arithmetical function. There are, in general, three types of computing shift registers, the differences being concerned with the behaviour of the extremes of the register:

(a) *Logical shifts* inject 0s at the bottom when shifting up and 0s at the top when shifting down. 11010 shifted down becomes 01101.
(b) *Arithmetic shifts* maintain the sign bit (MSB) when shifting up or down. 11010 shifted down becomes 11101. 01010 shifted down becomes 00101.
(c) *Cyclical shifts* link top and bottom bits. 10100 shifted up becomes 01001 (top bit reappears at the bottom).

7.6.7 Chain code generator

A chain code generator is used in data transmission where digital information is to be passed in serial form over a noisy channel. It converts, say, a four-bit parallel data word into a unique 15-bit serial string. The unique 15-bit string allows the original four-bit data to be deduced even if the message is considerably corrupted by noise.

The basis of a four-bit chain code generator is shown on fig. 7.43. Data is parallel loaded into the shift register, then clocked out by 15 clock pulses. Each output code is unique. For example:

1101 gives 110 1011 1100 0100
and 1111 gives 111 1000 1001 1010

The generator is sometimes called a pseudo-random generator.

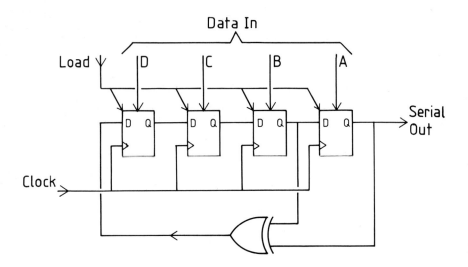

Fig. 7.43 Chain code generator.

7.7 Counters and shift registers

7.7.1 TTL
Clock figures are typical. Guaranteed speeds are roughly half.

Number	Description	Type	Clock MHz
7493	Four-bit binary ripple with reset	N	32↓
	—	L	3↓
	—	LS	32↓
74161	Four-bit synchronous with load. Direct clear	N	25↑
	—	LS	25↑
74163	Four-bit synchronous with synchronous clear	N	25↑
	—	LS	25↑
	—	S	40↑
74169	Four-bit synchronous up/down with load	S	40↑
74177	Four-bit ripple with load	N	35↓

Number	Description	Type	Clock MHz
74191	Four-bit reversible synchronous with load	N	20↑
	—	LS	20↑
74193	Four-bit synchronous up/down dual clock with load	N	25↑
	—	L	3↑
	—	LS	25↑
74197	Four-bit ripple with load	N	50↓
	—	LS	30↓
	—	S	100↓
74293	Four-bit ripple with reset	N	32↓
	—	LS	32↓
74393	Dual four-bit ripple with reset	N	25↓
	—	LS	25↓
74669	Four-bit synchronous up/down with load	LS	25↑

Decade counters

7490	BCD ripple counter set 0, 9	N	32↓
	—	L	3↓
	—	LS	32↓
74160	Synchronous BCD preset direct clear	N	25↑
	—	LS	25↑
74162	Synchronous BCD, preset synchronous clear	N	25↑
	—	LS	25↑
	—	S	40↑
74168	BCD up/down synchronous with load	S	40↑
74176	BCD ripple counter with load	N	35↓
74190	Synchronous BCD up/down with load	N	20↑
	—	LS	20↑
74192	Synchronous BCD up/down dual clock	N	25↑
	—	L	3↑
	—	LS	25↑
74196	BCD ripple counter with load	N	50↓
	—	LS	30↓
	—	S	100↓

Number	Description	Type	Clock MHz
74290	BCD ripple counter	N	25↓
	—	LS	32↓
74390	Dual BCD ripple counter	N	25↓
	—	LS	25↓
74490	Dual BCD ripple counter set 0, 9	N	25↓
	—	LS	25↓
74668	Synchronous BCD up/down with load	LS	25↑

Miscellaneous counters

Number	Description	Type	Clock MHz
7492	Divide by 12 (2, 6) with reset	N, LS	32↓
74142	BCD counter/latch/decimal decoder/driver	N	20↑
74143	BCD counter, seven-segment decoder	N	12↑
74144	BCD counter, seven-segment decoder	N	12↑

Shift registers

Number	Description	Type	Clock MHz
7491	Eight-bit serial in, serial out gated input	N	10↑
	—	L	3↑
	—	LS	10↑
7494	Four-bit dual presets	N	10↑
7495	Four-bit bidirectional parallel load	N	25↓
	—	L	3↓
	—	LS	25↓
7496	Five-bit parallel load	N	10↑
	—	L	5↑
	—	LS	25↑
7499	Four-bit bidirectional parallel load	L	3↓
74164	Eight-bit with clear	N	25↑
	—	L	12↑
	—	LS	25↑
74165	Eight-bit parallel load serial out	N	20↑
	—	LS	25↑
74166	Eight-bit parallel load serial out	N	25↑
	—	LS	25↑
74178	Four-bit parallel load	N	25↓
74179	Four-bit parallel load plus clear	N	25↓
74194	Four-bit bidirectional load	N	25↑
	—	LS	25↑
	—	S	70↓

Number	Description	Type	Clock MHz
74198	Eight-bit bidirectional parallel load	N	25↑
74199	Eight-bit bidirectional parallel load	N	25↑
74295	Four-bit bidirectional parallel load	LS	25↓
74299	Eight-bit bidirectional parallel load tristate output	LS	35↑
	—	S	50↑
74322	Eight-bit parallel load tristate output	LS	35↑
74323	Eight-bit bidirectional parallel load tristate output	LS	35↑
74395	Four-bit parallel load tristate output	LS	25↓

7.7.2 CMOS

Clock figures quoted are typical at 5 volts; 10 and 15 volt speeds are roughly double and triple. Guaranteed speed is half the above speeds.

Number	Description	Clock MHz
Binary counters		
4020	14-stage binary ripple	10↓
4022	Divide-by-8 Johnson	6↓ & ↑
4024	Seven-stage binary ripple	10↓
4029	Synchronous up/down binary/BCD	8↑
4040	12-stage binary ripple	10↓
4060	14-stage binary with oscillator	6—
4516	Binary up/down	10↑
4520	Dual four-bit binary	6↓ & ↑
4521	24-stage freq divider	12—
4526	Programmable four-bit down counter	12↓ & ↑
40161	Four-bit synchronous, asynchronous reset	10↑
40163	Four-bit synchronous, asynchronous reset	10↑
40193	Four-bit up/down. Separate clocks	5↑
Decimal counters		
4017	Five-stage Johnson counter, decoder outputs	6↓
4029	Synchronous up/down binary/BCD	8↑
4510	BCD up/down with load	10↑
4518	Dual BCD	6↓ & ↑
4522	Programmable BCD down	12↓ & ↑
40160	Synchronous BCD asynchronous reset	10↑
40162	Synchronous BCD synchronous reset	10↑
40192	BCD up/down separate clocks	5↑

Number	Description	Clock MHz
Miscellaneous counters		
4018	Presettable divide by N	4↑
4059	Programmable divide by N	8−
4751	Universal divider	5−
Shift registers		
4006	18-stage arranged 2×5, 2×4 SISO	18↓
4014	Eight-bit parallel in, serial out	9↑
4015	Dual four-bit serial in, parallel out	9↑
4021	Eight-bit parallel in, serial out	9↑
4031	64-stage serial in, serial out	5↑
4035	Four-bit parallel in/out	10↑
4094	Eight-stage serial in, parallel out tristate	10↑
4517	Dual 64-bit serial in/out	5↑
4557	64-bit variable length serial in/out	5↓ & ↑
4731	Quad 64-bit serial in/out	6↓
40194	Four-bit bidirectional parallel in/out	12↑
40195	Four-bit parallel in/out	10↑

CHAPTER 8

Arithmetic Circuits

8.1 Introduction

The theory behind binary and BCD arithmetic was given in chapter 6. In this chapter the theory is translated into practical circuits. These circuits are the basis of many logic systems from direct digital control to computers.

8.2 Encoders and decoders

8.2.1 Decimal to binary encoding

Where a circuit interfaces with the outside world, the incoming signals are often in decimal form. In all bar the simplest applications, a BCD or binary representation is required before arithmetic operations can be performed.

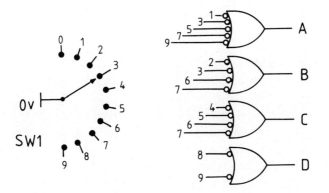

Fig. 8.1 Representation of decimal to binary encoder.

Usually, decimal inputs are provided with ten lines per decade, one line being taken high (or low) at a time. For simplicity, we shall consider one decade being supplied from a ten-position rotary switch. The encoding of one decade into binary form is a simple combinational logic problem. One possible solution, where one line at a time is pulled low, is shown in fig. 8.1. The negative NOR gates could, of course, be implemented by positive NAND gates as explained in section 3.2.8. The circuit gives a true BCD output. Note that the 0 line is not required by the encoder.

MSI devices called 'encoders' are not available, but devices called 'priority encoders' fulfil the same function. These devices take a decimal input on nine lines (as fig. 8.1) and give the corresponding BCD output. Unlike fig. 8.1, however, a priority encoder can accept more than one input signal line in an active state at once. In these conditions, the priority encoder indicates the *highest* signal present. If, for example, inputs 1, 2, 6 and 8 were all present, a priority encoder would give 1000, the BCD equivalent of 8, this being the highest input state.

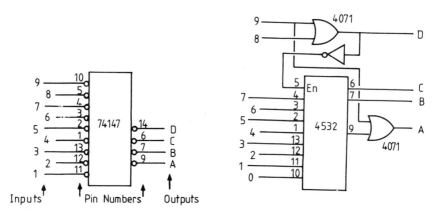

Fig. 8.2 Decimal encoders using priority encoder ICs. (a) TTL 10-line encoder. (b) CMOS 10-line encoder using octal 4532.

The TTL family has a useful 10-line priority encoder in the 74147 which can convert low-going decimal inputs to complementary binary in one device. The equivalent CMOS priority encoder is the 4523. Unfortunately this device is only an eight-line encoder, so additional logic is required to give true decimal to BCD conversion. Comparison of the 74147 and the 4523 is given in fig. 8.2.

Where inputs are being obtained from rotary switches, it should be observed that encoding thumbwheel switches are available. These give true BCD contact closures on four outputs, and dispense with the need for an encoding stage.

Multi-decade inputs, when encoded, give multi-decade BCD. If true binary representation is required, further encoding is needed. This slightly more complex conversion is discussed further in section 8.10.

8.2.2 *Binary to decimal decoding*
Four binary bits can represent from 0 to 15 decimal, so it is first necessary to distinguish between BCD (0-9) and binary decoding. BCD, being simpler, will be dealt with first.

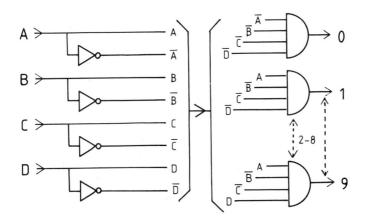

Fig. 8.3 Binary (BCD) to decimal encoder.

BCD to decimal conversion is, again, straightforward combinational logic, the principle of which is shown in fig. 8.3. Nine four-input AND gates plus four inverters are required. In practice, it is almost always more convenient to use an MSI decoder.

TTL, in particular, has a wide range of decoders with logic outputs (7442) or open collector outputs, with a variety of voltage and current ratings (7445, 74141, 74145). The CMOS equivalent is the 4028 decoder.

XS3 code was described in section 6.9.3. Decoders from XS3, and XS3 Gray code to decimal are also available in MSI form in the TTL family. These are, respectively, the 7443 and 7444.

Full binary decoding is also a simple combinational logic problem. A discrete decoder requires sixteen four-input AND gates and four inverters. TTL has one full binary decoder, the 74154, with TTL outputs. If open collector outputs are required, two BCD decoders can be cascaded as in fig. 8.4. CMOS binary decoders (4514, 4515) incorporate four-bit latches on the binary inputs. The 4514 gives true decimal outputs, the 4515 complement outputs.

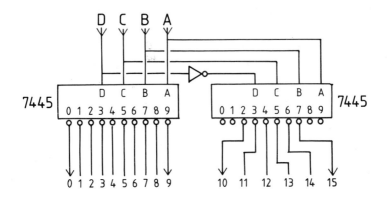

Fig. 8.4 Full binary decoding with two BCD decoders.

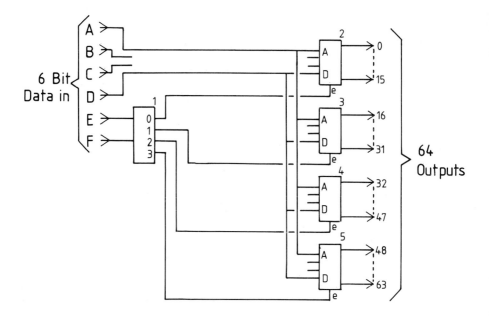

Fig. 8.5 Cascaded decoders giving 64 decoded outputs from a six-bit input.

A binary output is often required to drive some form of display. This is a special form of decoding and is discussed further in chapter 9. Decoding of large binary numbers to multi-digit BCD is a more complex conversion, and is described in section 8.10.

Although not strictly binary to decimal decoding, a decoding circuit may be required to give, say, 64 decoded outputs from a six-bit binary number. This can be achieved by cascading smaller decoders, and selecting a decoder via the enable inputs and a decoder working on the two most significant bits as shown in fig. 8.5. IC4, for example, provides outputs 32-47.

8.3 Binary addition

The theory of binary arithmetic was given in section 6.4. The basic arithmetic circuit is the single-bit adder, which can be represented by the block diagram of fig. 8.6a. This has three inputs; the two binary bits to be added, A, B, and a carry input from the preceding (less significant) stage. There are two outputs; the actual sum, and the carry to the succeeding (more significant stage).

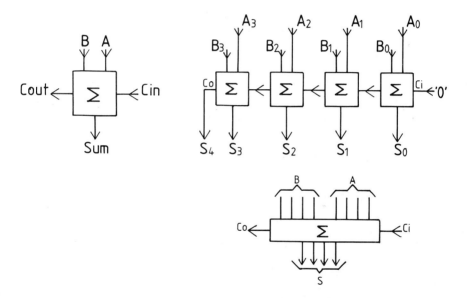

Fig. 8.6 Simple adders. (a) Full adder. (b) Four-bit adder. (c) General symbol.

With the adder block of fig. 8.6a, an adder of any required size can be produced. Figure 8.6b shows a four-bit adder using four adder blocks. Note that the addition of two four-bit numbers produces a five-bit result with the MSB carry out providing the fifth bit. Note also that the LSB carry in is connected to a 0.

Figure 8.6b can be represented by fig. 8.6c, which is the conventional block for an adder. It is usual to draw arithmetic circuits with normal binary representation, i.e. LSB to right, MSB to left.

The adder block of fig. 8.6a has the truth table:

Cin	B	A	Sum	Co
0	0	0	0	0
0	0	1	1	0
0	1	0	1	0
0	1	1	0	1
1	0	0	1	0
1	0	1	0	1
1	1	0	0	1
1	1	1	1	1

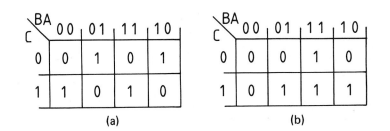

(a) (b)

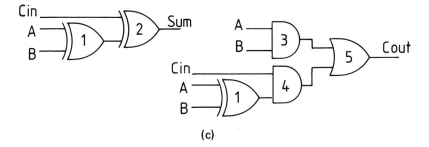

(c)

Fig. 8.7 The full adder. (a) Sum Karnaugh map. (b) Carry Karnaugh map. (c) Sum and carry circuit.

The adder is therefore two simple combinational logic circuits; each with three inputs. The Karnaugh maps are shown on fig. 8.7a, b. The sum map suggests that a four-term solution is required, but as explained in section 3.3.6, this type of chequer-board pattern can be obtained with XOR gates. The corresponding circuits are shown on fig. 8.7c. Separate sum and carry circuits are shown for simplicity, but examination will show that gates 1 are identical. A full adder requires just five two-input gates.

A full adder can also be derived by considering the so-called half adder circuit of fig. 8.8a. This has just AB inputs (no carry in) and the usual sum and carry outputs. A full adder is constructed from two half adders as in fig. 8.8b. The sum output from AB is added to the carry in, and the two carry outs are ORd. A half adder has the simple truth table:

B	A	Sum	Carry
0	0	0	0
0	1	1	0
1	0	1	0
1	1	0	1

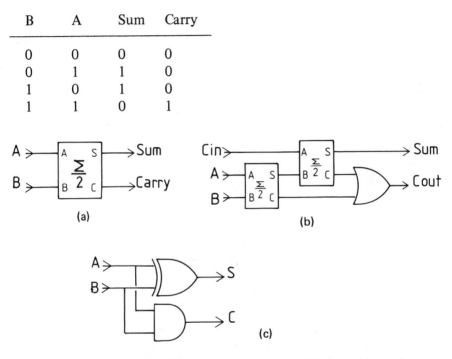

Fig. 8.8 The half adder. (a) Half adder block diagram. (b) Full adder with two half adders. (c) Circuit diagram.

The sum is produced directly by an XOR gate and the carry by an AND gate, so the circuit of fig. 8.8c can be produced without resorting to a Karnaugh map.

8.4 Serial/parallel arithmetic

The adder of figs 8.6b and c is called a parallel adder, because the addition of all four bits is produced in parallel, or simultaneously. Neglecting propagation delays, parallel addition is an instantaneous process.

It is also possible to perform addition one digit at a time. Such an operation is called serial addition, and is shown diagrammatically in fig. 8.9. The circuits consists of three shift registers, SRA, B, C, one single-bit adder and one D-type flip-flop.

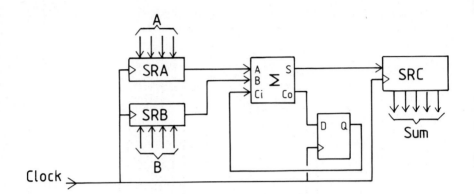

Fig. 8.9 Four-bit serial adder.

At the start of the operation, the two numbers to be added are parallel-loaded into shift registers A and B, and the D-type is cleared. The adder outputs, at this point, will be the result of adding the LSBs of numbers A and B. On receipt of the first clock pulse, the sum output will be shifted into C, the carry output clocked into the D-type, and the next LSBs brought to the adder inputs AB. These are added with the carry from the previous bit held on the D-type.

As further clock pulses arrive, each pair of bits will be added in turn with the carry from the previous bit, and the results shifted into shift register C. When all bits have been added, the final carry will be transferred into shift register C, whose length is therefore one bit longer than A, B (as required by the rules of arithmetic). Register C now contains the result of the addition of A and B.

Serial addition is economical in terms of devices, but is relatively slow by comparison with parallel addition. To add two N-bit words, N + 1 clock pulses are required. Usually, a serial adder can only be used in a synchronous system.

Serial arithmetic is not only limited to addition. It is equally easy to make a serial subtractor, and serial methods of conversion between large binary and BCD numbers are described in section 8.10.1. Serial techniques of data transmission are discussed in chapter 11.

8.5 High speed adders and look ahead carries

Consider the simple binary addition:

```
 11111111
 00000001
100000000
```

The addition of the two 1s in the LSBs produces a carry which must propagate through all the more significant stages to produce the MSB 1.

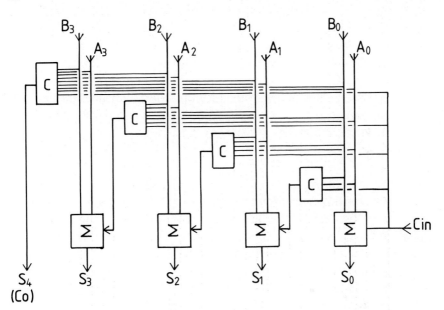

Fig. 8.10 Look ahead carry.

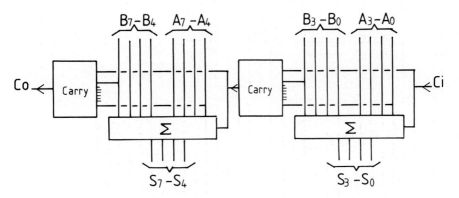

Fig. 8.11 Look ahead carry across four bits (as used on 7483).

In simple parallel adders such as fig. 8.6b, this rippling through of the carry is the major restriction on speed. This restriction can be overcome by a technique called a 'look ahead carry'.

The fundamental principles are illustrated by the fast four-bit adder of fig. 8.10. At each stage, the carry in is generated not by the preceding LSB stage, but by totally separate combinational logic circuits (c) which look at the least significant bits. Because each carry is generated simultaneously, there is no cumulation of propagation delays.

Figure 8.10 shows that the complexity of the look ahead carry circuit increases dramatically with the number of bits to be added. In many circuits (and most MSI adders) a reasonable compromise is achieved by using look aheads over blocks of, say four bits, as shown in fig. 8.11. Here, two numbers are added with ripple carries being used between blocks of four bits, and look ahead circuits used between blocks.

8.6 MSI adders

It is usually more convenient, of course, to use MSI devices to construct arithmetic circuits. Typical of these is the CMOS 4008 shown in fig. 8.12. This is a four-bit adder, using the look ahead carry across four bits idea of fig. 8.11. Several devices can be cascaded to give an adder of any desired length, the limit being the cumulative carry propagation of 200 ns per device.

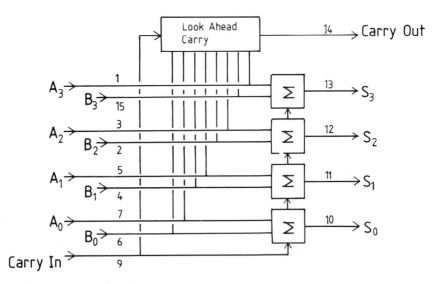

Fig. 8.12 4008 four-bit adder with across-stage look ahead carry.

The TTL equivalent is the 7483 (and the improved version 74283 with standard TTL supply pins). Although the 7483 and 74283 are functionally identical to the 4008, they are considerably faster. The normal version has a carry propagation delay of 10 ns per device, with the Schottky version faster still at 7.5 ns. Using the latter, two 16-bit words can be added in 30 ns. ECL arithmetic circuits are even faster. Standard ECL adders have carry propagation delays as low as 2 ns.

There are other adders in the TTL family. The 7480 provides a single-bit adder, and the 7482 a two-bit adder. A dual one-bit adder with high fan-out is available in the 74183, this latter device being designed for use with multiplier circuits described in section 8.9. In general, however, the designer will use the 4008 in CMOS designs and the 74283 with TTL.

In section 8.13 more complex LSI chips will be described.

8.7 Subtraction circuits

There are no MSI 'subtractors' as such because subtraction can be obtained by complementation followed by addition as described in section 6.6. To subtract 3 from 9, for example, we first obtain twos complement:

    0011   3 represented in binary
    1100   complement
    1101   twos complement of 3

Now perform addition with 9 and twos complement 3:

    1001   9
    1101   twos complement 3
    0110   6 (top carry discarded)

which is, of course, the correct answer.

The principle is easily translated into a practical circuit. Figure 8.13 shows an eight-bit subtractor using the CMOS 4008. The circuit produces $(A - B)$. Complementation of B is achieved with eight inverters (two 4049 hex inverters), and the addition of 1 to give twos complement is obtained by forcing a 1 into the LSB carry in. Note that the carry out is not used in the MSB adder. The circuit could be implemented in TTL with the 74283 adder and 7404 inverters.

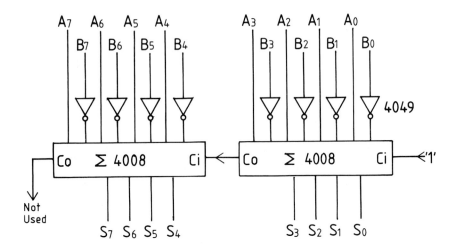

Fig. 8.13 Eight-bit subtractor.

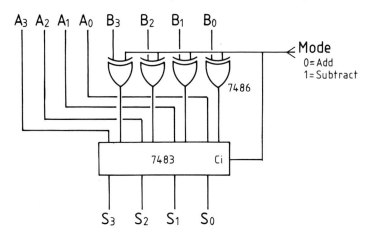

Fig. 8.14 Basic adder/subtractor circuit.

An arithmetic unit which can add or subtract is shown in fig. 8.14. Selectable inversion is obtained with XOR gates. When the mode control line is 1, the XOR gates invert the B inputs, and a 1 is forced into the LSB carry in, giving an output $(A - B)$. When the mode control is 0, B is not inverted and the LSB carry in is 0. The output is accordingly $A + B$. (The action of the XOR gate is described in section 3.2.9.)

If the two numbers are in twos complement form, the circuit must be modified to keep the correct sign. A twos complement adder/

subtractor, shown in fig. 8.15, will correctly perform sums such as:

$$8 + (-3) = 5$$
$$-5 + (-2) = -7$$
$$3 - (-5) = 8$$
$$2 - 7 = -5$$

giving the correct signed answer in each case. Note that the B sign bit is not connected via an XOR gate.

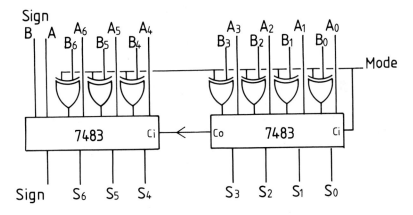

Fig. 8.15 Signed twos complement adder/subtractor.

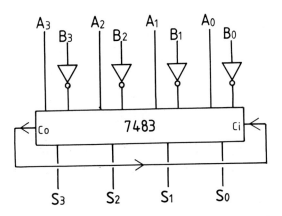

Fig. 8.16 Ones complement subtractor with end around carry.

The ones complement representation and the use of the end around carry was described in section 6.6. Using these principles it is

224

possible to build a subtractor, an example of which is shown in fig. 8.16. The end around carry is produced by linking carry out to carry in.

Multipurpose arithmetic chips are described in section 8.13.

8.8 Comparators

Comparator circuits are required where the magnitudes of two binary numbers are to be compared. Typical applications are device address decoding, and digital position controls. A comparator can be represented by fig. 8.17a, and has, as inputs, two binary numbers A, B. The three outputs show the result of comparing A and B giving $(A < B)$, $(A = B)$, $(A > B)$. By combining these with an OR gate expressions such as $A = > B$ can be obtained.

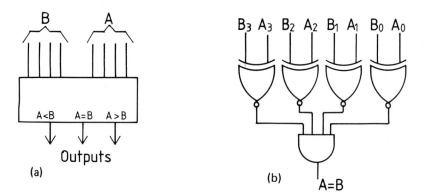

Fig. 8.17 Four-bit comparator. (a) Comparator symbol. (b) Circuit for $A = B$.

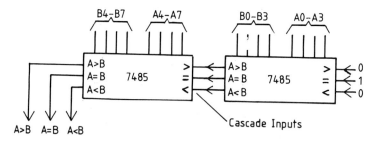

Fig. 8.18 Cascading comparators.

Provision of the three outputs is a simple combinational logic problem. The $<$ and $>$ outputs are obtained from the MSB carries of subtraction circuits, and the equality output by extension of the four-bit comparator of fig. 8.17b. In practice, however, it is usually simpler to use an MSI device.

Typical of these is the TTL 7485 four-bit comparator. This compares two four-bit numbers, and gives the $< = >$ outputs described above. Three inputs corresponding to $< = >$ are also provided to allow several devices to be cascaded as in fig. 8.18. Note that the cascading is from LSB to MSB, with outputs from the most significant device. It would be feasible to design the cascading from MSB to LSB, but most devices use the scheme as shown.

The CMOS four-bit comparator 4585 operates identically to the 7485, and again cascades from LSB to MSB, with outputs being taken from the most significant device.

Both the 4585 and 7485 are described as four-bit binary comparators, but will work equally well with all common BCD codes (including XS3). The two 8-bit numbers on fig. 8.18 could be two-decade BCD numbers, and the three outputs would indicate the relative magnitudes correctly.

8.9 Multiplication and division

8.9.1 Introduction

Multiplication is rarely required in logic circuits, and the necessary circuits are somewhat complex. If a system requires a considerable number of multiplication operations it is usually worth considering some form of microprocessor-based assembly. The circuit will be relatively cheap and simple, and multiplication can be obtained by standard program subroutines.

If the hardware multiplication is required there are three basic techniques. All are based on the principles outlined in section 6.7. (It should, of course, be observed that multiplications by 2, 4, 8, 16, etc. can be obtained by a simple shift register). Rate multipliers (described in section 8.12) can perform multiplication and division, albeit slowly.

8.9.2 Parallel addition

It was shown in section 6.7 that binary multiplication is analogous to decimal multiplication. The result is obtained by the addition of weighted partial sums, each partial sum being obtained by a parallel AND of the multiplier and one bit of the multiplicand.

The circuit shown in fig. 8.19 multiplies a four-bit number (A) by a three-bit number (B) to give a seven-bit result. The partial sums are obtained by gating A by each bit of B. The partial sums are weighted by position and added in succession to give the final result. Figure 8.19 corresponds almost directly to the pen and paper method.

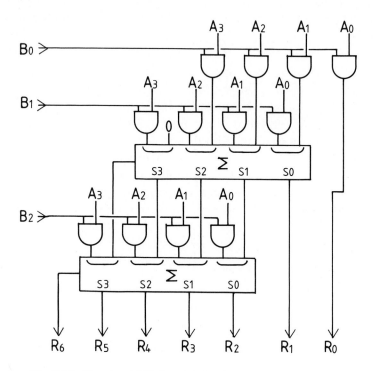

Fig. 8.19 Circuit of 3 × 4 multiplier (to give seven-bit result).

Although relatively simple where small numbers are to be multiplied, the circuit complexity (and propagation delays) increase dramatically with the number of bits.

8.9.3 Shift and add multiplier
A larger (but slower) multiplier can be constructed as a sequential circuit using a single adder, and a shift register to give the positional weighting.

An 8 × 8 bit multiplier using this principle is shown in block diagram form in fig. 8.20. Number A is loaded into the least significant end of a 16-bit shift register; number B into an eight-bit latch. Associated with B is an eight-way multiplexer which selects one bit of B at a time.

At the start of the operation, the 16-bit result register R is cleared. A is gated with the LSB of B and added to R to give a partial result which is latched back into R. A is now shifted up one bit and the next LSB of B selected by the multiplexer. The gated version of A, correctly positioned, is added again to R to give a new partial result. The operation is repeated until, after eight shifts and adds, R will contain the 16-bit result of A × B. The operation is therefore rather slow.

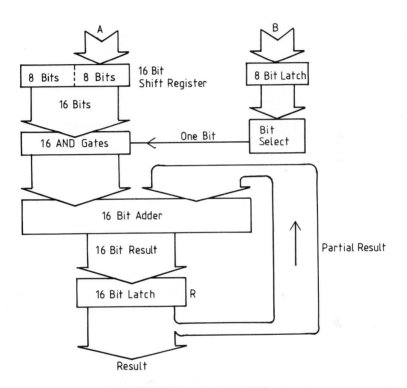

Fig. 8.20 Shift and add muliplier.

The complexity of fig. 8.20 does not increase with the size of the numbers to be multiplied. A 16 × 16 multiplier would have the same layout, but the shift register, AND gates, adder and R would be 32 bits in length, and the B latch and multiplexer would be 16 bits in length. Sixteen shift and add operations would be required, so the circuit would operate at half the speed of an 8 × 8 multiplier.

8.9.4 Look-up tables
The fastest multiplication circuits are based on ROMs. The multiplier and multiplicand are, together, used to address one location in the ROM where the result can be looked up. Two 1K bit ROMs can form

a 4 × 4 multiplier with no additional devices.

The TTL family has two suitably programmed ROMs in the 74285 and 74285 devices. These give, respectively, the most significant and least significant four-bit 'nibbles' of a 4 × 4 multiplication and are connected as in fig. 8.21. The circuit propagation delay is 40 ns.

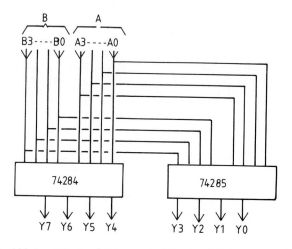

Fig. 8.21 4 × 4 multiplier (giving eight-bit result) using look-up PROMs.

Larger multipliers are built using a scheme called a Wallace tree. The theory is best illustrated by a decimal example. Let us suppose we wish to multiply two four-digit decimal numbers, and we have a 'building block' that can multiply two two-digit decimal numbers. For example, let us multiply 4057 by 8753. Using our 2 × 2 building block we form four partial sums:

$$40 \times 53 = 2120$$
$$40 \times 87 = 3480$$
$$57 \times 53 = 3021$$
$$57 \times 87 = 4959$$

These partial results are now added, suitably positionally weighted, to give the final result.

40 × 87	34 800 000
40 × 53	212 000
57 × 87	495 900
57 × 53	3 021
	35 510 921

The formation of the partial results and their addition is performed by a circuit called a Wallace tree.

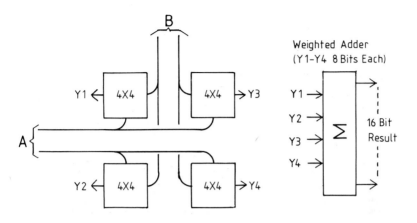

Fig. 8.22 Block diagram of fast 8 × 8 multiplier (giving 16-bit result).

A fast binary 8 × 8 multiplier is shown diagrammatically in fig. 8.22. Four eight-bit partial sums Y1 to Y4 are formed by four 4 × 4 multipliers, each identical to fig. 8.21. These are added to give the 16-bit result.

The circuit of fig. 8.22 is fast (typically 70 ns) but not particularly simple. Sixteen MSI packages are required (4 off each 74284, 74285, 74183 adders, 3 off each 74181 adders and one 74182 look ahead carry). The circuit complexity increases markedly with the number of bits. A 16 × 16 multiplier requires over 70 MSI devices but is still very fast at 100 ns.

It is possible to reduce the circuit complexity with larger ROMs, but such devices are not readily available pre-programmed with multiplication look-up tables. A designer with a large volume production product would probably find it economic to have purpose-made devices.

8.9.5 *Signed multiplication*
Multiplication of signed binary numbers is a rather daunting task. Usually the easiest solution is to convert all numbers to a positive form, perform the multiplication, then apply the correct sign using the relationships:

(+) X (+) = (+)
(+) X (−) = (−)
(−) X (−) = (+)

For example 3 × (−7) = 3 × 7 = 21, sign −, result −21. Or (−5) × (−3) = 5 × 3 = 15, sign +, result +15.

Such operations can be performed on twos complement numbers, but the circuits are necessarily complex.

8.9.6 Division

The operation of division is even more difficult than that of multiplication, and is rarely needed. It is possible to use a shift/subtraction variation on the circuit of fig. 8.20, but such circuits are slow. Where fast division is required, a microprocessor-based system or an LSI arithmetic chip (see section 8.13) is probably the best solution.

8.10 BCD conversion and arithmetic

8.10.1 Conversion

Simple binary/BCD conversion was described in section 8.2. For multi-digit numbers, the conversion becomes unduly unwieldly if straightforward combinational logic is used.

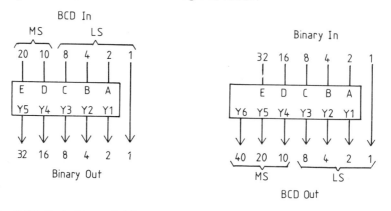

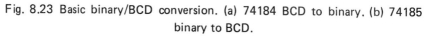

Fig. 8.23 Basic binary/BCD conversion. (a) 74184 BCD to binary. (b) 74185 binary to BCD.

Conversion between binary and BCD can be done via look-up tables stored in ROMs. Two such devices are the TTL 74184 (BCD to binary) and 74185 (binary to BCD) converter. The basic devices are shown in fig. 8.23.

The 74184 takes a six-bit BCD number (up to decimal 39) and gives the corresponding binary output. The 74185 takes a six-bit binary number (up to decimal 63) and gives a BCD output. Note that in each case the LSB bypasses the converter.

These devices can, however, be cascaded to convert numbers of any desired length. The steps for a full eight-bit BCD number are:

BCD number $(10 \times A) + B$ where A, B are BCD

$$= 10 (A_3 2^3 + A_2 2^2 + A_1 2 + A_0) + B$$

where A_n are bits (for example, for BCD 57, $A_3 = 0$, $A_2 = 1$, $A_1 = 0$, $A_0 = 1$, $B = 0111$)

Rearranging gives:

$$10 (A_3 2^3 + A_2 2^2) + (10(A_1 2 + A_0) + B)$$

The right bracket can be converted by a 74184 giving

$$10 (A_3 2^3 + A_2 2^2) + (C_5 2^5 + C_4 2^4 + C_3 2^3 + C_2 2^2 + C_1 2 + C_0)$$

Regrouping gives:

$$2^2 [10 (A_3 2 + A_2) + C_5 2^3 + C_4 2^2 + C_3 2 + C_2] + C_1 2 + C_0$$

The left bracket can be converted by a 74184, and is positioned two bits to the left to implement the multiplication by 4. The resulting circuit is shown in fig. 8.24a.

Similar arithmetic solutions give the circuits for any required BCD to binary conversion, but it is usually easier to look up the application notes. A three-decade conversion is shown in fig. 8.24b.

Multi-digit binary to BCD conversion is obtained in a similar way (and it is again easier to look up device application notes than to design a circuit from scratch). An eight-bit binary to BCD conversion is shown in fig. 8.25.

Conversion can also be performed with counters, and a simple, but slow, converter is shown in fig. 8.26. A BCD number is parallel-loaded into a BCD down converter. Flip-flop 1 is then set. Clock pulses are then gated to the BCD down counter and a suitable length binary up counter. When the BCD counter reaches zero, gate 2 resets the flip-flop and stops further pulses. The binary counter now contains the binary equivalent of the number loaded into the BCD counter. Obviously, the reverse conversion can be obtained between binary and BCD by using a binary down counter and a BCD up counter.

A straightforward conversion for medium-sized numbers can be

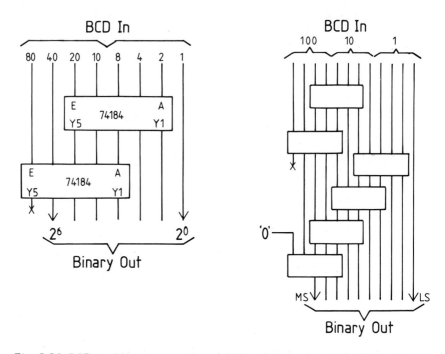

Fig. 8.24 BCD to binary conversion. (a) Two-decade circuit. (b) Three-decade circuit.

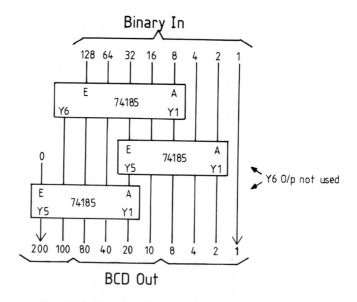

Fig. 8.25 Eight-bit binary to BCD conversion.

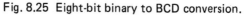

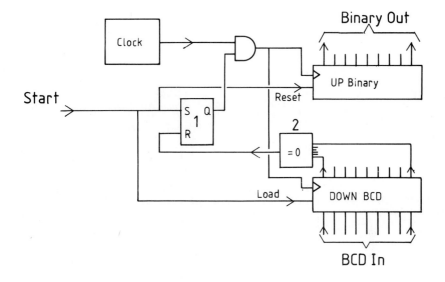

Fig. 8.26 Counter BCD to binary converter.

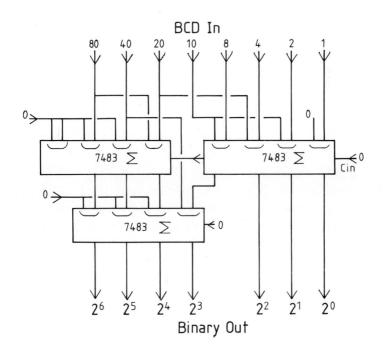

Fig. 8.27 BCD to binary conversion using adders.

234

obtained by converting each BCD bit to its binary equivalent (e.g. 80 = 64 + 16, 40 = 32 + 8) and adding the results with the binary adders. A two-decade BCD to binary converter using this technique is shown in fig. 8.27.

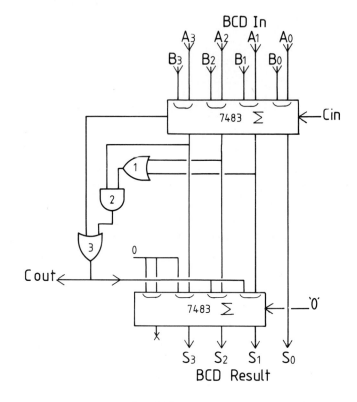

Fig. 8.28 BCD adder.

8.10.2 BCD arithmetic

The principle of BCD arithmetic was outlined in section 6.9.2. Essentially, each decade is dealt with individually, and arithmetic is performed in binary. The result is then examined and a correction factor of 6 is applied if the result is a non-valid BCD result.

A typical BCD adder (for one decade) is shown in fig. 8.28. The two BCD numbers (A, B) are added by a standard 7483 *binary* adder. Gates 1, 2 decode non-valid BCD results (1010, 1011, 1100, 1101, 1110, 1111). If one of these results occurs *or* a binary carry is produced, 6 is added to the result by a second adder to give a correct BCD result. OR gate 3 provides the carry to the next decade. Adders of any required length can be constructed.

Subtraction is similar, but a correction factor of 6 is subtracted if a carry is generated or a non-valid code produced. Figure 8.28 will produce $(A - B)$ if inverters are added to the B inputs and between gate 3 output and the correction factor inputs to the second adder.

If BCD multiplication or division is required, it is probably easier to convert the number to binary, perform the operation in binary and reconvert the result.

8.11 Code conversion

Conversion between other codes and binary is often required. A common conversion is between Gray code (possibly from some position encoder) and binary. Gray to binary conversion (and the inverse operation) are shown in fig. 8.29.

Generalised code conversion can be represented by fig. 8.30. An M-bit input code is to be converted to an N-bit output code. Any conversion can be achieved by considering the circuit as N combinational logic circuits as in fig. 8.30b, which are designed using the techniques described in chapter 3. The design can usually be simplified by looking for common terms in the N combinational circuits.

8.12 Rate multipliers

The rate multiplier is an often neglected device which can provide many arithmetic functions in applications where speed is not important. The basic circuit is shown in fig. 8.31, and consists of a three-stage ripple counter whose outputs are gated by the three control inputs X0, X1, X2.

Waveforms at various points in the circuit are shown on fig. 8.32. It can be seen that with $X0 = 1$, the output frequency is one-half of the input frequency, with $X1 = 1$ one quarter, and with $X2 = 1$, one eighth. If more than one input is 1, the output frequency will be the sum of the individual frequencies. If, for example, $X0 = 1$, $X1 = 0$, $X2 = 1$ the output frequency will be 5/8 of the input frequency. The parallel input can be considered as a binary fraction 0.101. Note that the output frequency is *not* symmetrical. Rate multipliers work with pulses counted over some time period.

The parallel input to the rate multiplier can also be considered as an integer N, when the output frequency is given by:

$$Fo = \frac{N\, Fin}{R}$$

236

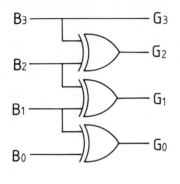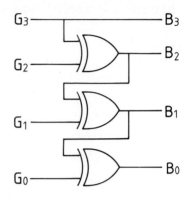

Fig. 8.29 Binary/Gray conversion. (a) Binary to Gray. (b) Gray to binary.

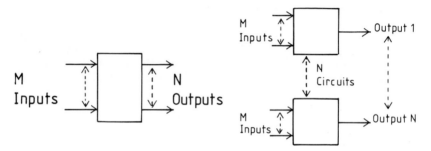

Fig. 8.30 Generalised code conversion. (a) Generalised code conversion. (b) Realisation with N combinational logic circuits.

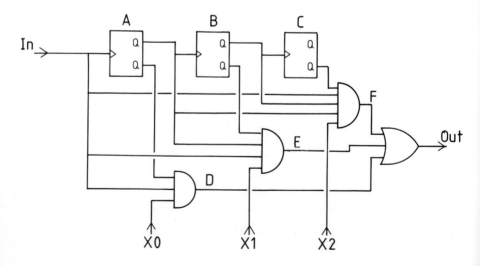

Fig. 8.31 Three-bit rate multiplier.

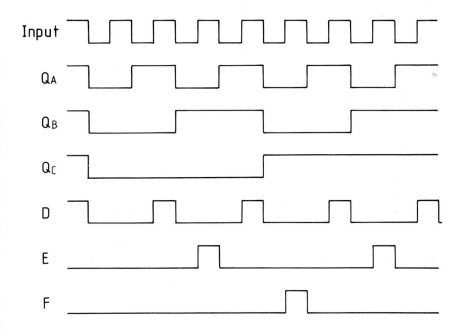

Fig. 8.32 Rate multiplier waveforms.

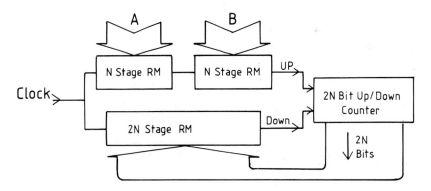

Fig. 8.33 Rate multiplier multiplication circuit.

where $R = 2^n$ for n stages (8 above).

Rate multipliers can be cascaded to any desired length. Figure 8.33 shows a typical rate multiplier circuit, which multiplies two n-bit binary numbers A, B. The numbers A, B are inputs to two n-stage rate multipliers, whose output frequency will be:

$$f_2 = \frac{A.B}{R} f_1$$

where $R = 2^{2n}$.

This frequency is connected to the count up input of a 2n-stage counter. The counter outputs C are connected as inputs to a second 2n-stage rate multiplier, whose output frequency will be:

$$f_3 = \frac{C}{R} f_1$$

where $R = 2^{2n}$ as above.

At balance, the counter output will be stable when $f_2 = f_3$ or

$$\frac{C}{R} f_1 = \frac{AB}{R} f_1$$

or C = AB
The circuit will stabilise with C containing a binary number equivalent to the product of AB.

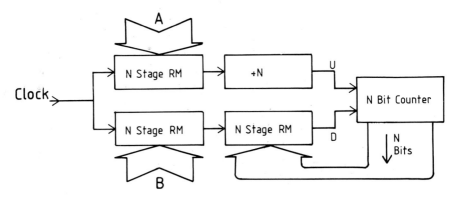

Fig. 8.34 Division using rate multipliers.

A division circuit is shown in fig. 8.34. Using similar reasoning to that above:

$$f_2 = \frac{Af_1}{2^n.2^n} = \frac{Af_1}{R}$$

and

$$f_3 = \frac{Bf_1}{2^n} \times \frac{C}{2^n} = \frac{BCf_1}{R}$$

At balance

$$\frac{BCf_1}{R} = \frac{Af_1}{R}$$

or

$$C = \frac{A}{B}$$

Obviously, both circuits take time to stabilise, so the circuits are slow.

There are two rate multipliers available in the TTL family and one CMOS device. The 7497 is a six-bit binary rate multiplier, and the 74167 a four-bit BCD rate multiplier. The latter is really the only way to obtain true BCD multiplication and division without resorting to microprocessors. Both devices are synchronous, for high speed. The CMOS 4527 is also a four-bit BCD device.

8.13 LSI arithmetic devices

There are many LSI devices, designed primarily for the computer industry, which perform arithmetic operations. These devices often are not members of a specific logic family.

One of the simplest of these devices is the 74381 Arithmetic and Logic Unit (ALU). There are two four-bit binary inputs A, B and three mode selection inputs S0, S1, S2 which select the device operation as follows:

S2	S1	S0	Operation
0	0	0	Clear (outputs 0)
0	0	1	B − A
0	1	0	A − B
0	1	1	Arithmetic A + B (ADD)
1	0	0	$A \oplus B$
1	0	1	Logical A + B (OR)
1	1	0	AB (AND)
1	1	1	Preset (outputs 1)

Carry outputs and inputs are provided for cascading, and the device can be used with the 74182 look ahead carry generator. A more complex version of the ALU is the 74181 which has additional features including shifts, comparisons, NAND, NOR, plus other logic operations.

The most complex, and versatile, devices are designed for use with microprocessors. The AMD 9511, for example, is a 24-pin IC which can add, subtract, multiply and divide to 32 bits in fixed and floating point modes. Trigonometric functions, square roots, logarithms and exponentiation can also be performed. Data is loaded, functions selected and results output via a bidirectional data bus. The device is fast, 16-bit multiply or divide operations taking about 40 μs. Complex 32-bit trig functions take about 600 μs. Not surprisingly, the device is not cheap and at the time of writing one IC costs about as much as two portable televisions. There is also a simpler (and cheaper) device in the AMD 9512 which only performs the normal four arithmetic functions, albeit to 64-bit accuracy in fixed and floating point modes.

8.14 Arithmetic ICs

8.14.1 TTL

Adders/arithmetic and logic units (ALUs)

7480	One-bit adder	N
7482	Two-bit adder	N
7483	Four-bit adder	N, LS
74181	Four-bit ALU	N, LS, S
74182	Look ahead carry for 74181	N, S
74183	Dual one-bit adder	H, LS
74281	Four-bit accumulator	LS, S
74283	Four-bit adder (Improved 7483)	N, LS, S
74381	Four-bit ALU	LS, S
74382	Four-bit ALU	LS
74385	Quad serial adder/subtractor	LS
74681	Four-bit accumulator	LS

Code converters

74184	BCD to binary	N
74185	Binary to BCD	N

Decoders

7442	BCD to decimal	N, L, LS
7443	XS3 to decimal	N, L
7444	XS3 Gray to decimal	N, L

Other decoders with high voltage/high current outputs can be found with display devices, section 9.8.

Priority encoders

74147	BCD	N, LS
74148	Octal	N, LS
74278	Four-bit cascadable	N
74348	Octal with tristate	LS

Multipliers

74261	2 X 2 multiplier	LS
74274	4 X 4 multiplier	S
74275	Seven-bit slice, Wallace tree	LS, S
74284	4 X 4 multiplier	N
74285	4 X 4 multiplier	N
74384	Eight-bit serial/parallel twos complement multiplier	N

Rate multipliers

7497	Six-bit binary	N
74167	BCD	N

Miscellaneous

7485	Four-bit comparator	N, L, LS
7487	Four-bit true/complement/I/O gate	H

8.14.2 CMOS

Adders/Arithmetic and logic units (ALUs)

4008	Four-bit adder
4032	Triple serial adder true
4038	Triple serial adder complement
4560	BCD adder
4561	Nines complementer
4581	Four-bit ALU
4582	Look ahead carry

Decoders
4028   BCD to decimal

Priority encoders
4532   Eight-input priority encoder

Multipliers
4554   2 X 2 multiplier

Rate multiplier
4527   Rate multiplier

Miscellaneous
4585   Four-bit magnitude comparator

CHAPTER 9
Display Devices and Drivers

9.1 Introduction

Most digital systems incorporate an assortment of display devices, from simple on/off indicators to complex alphanumeric displays and VDUs. This chapter describes various display devices, their characteristics, requirements and the circuits to drive them.

9.2 Display devices

9.2.1 Light emitting diodes (LEDs)
An LED is (as its name implies) a fairly conventional diode which emits light when conducting. The symbol for an LED is shown in fig. 9.1a. When the anode is biased positive with respect to the cathode, current flows and light is emitted. When the anode is biased negative, a negligible leakage current flows and no light is emitted.

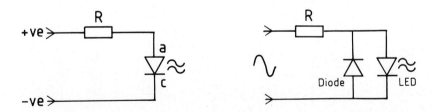

Fig. 9.1 The light emitting diode. (a) LED circuit. (b) Protection circuit.

An LED is a current operated device, and when conducting exhibits a fairly constant voltage drop of about 2 volts. A series resistor must always be used to define the current as shown on fig. 9.1a. Most LEDs operate on a current of between 5 mA and 20 mA. For an LED of forward drop Vf, a supply V and an LED current of I, the value of R is given by:

$$R = \frac{V - Vf}{I} \text{ ohms}$$

An LED cannot stand a reverse voltage in excess of 5 volts. If voltage reversal is possible, or an AC source is used, a protection diode as in fig. 9.1b must be used.

The light output from an LED is proportional to the current, but as the eye's response is logarithmic the apparent intensity is not greatly dependent on the current.

Fig. 9.2 Some LEDs.

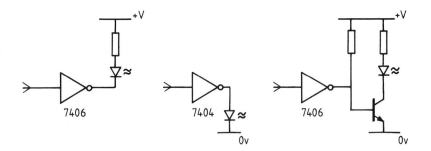

Fig. 9.3 Various circuits for driving LEDs with TTL. (a) On for 1 input. (b) On for 0 input. (c) High intensity.

LEDs are available in a range of sizes from 1.5 mm to over 5 mm (see fig. 9.2) and in red, green or yellow colours. Infra-red LEDs for photocell applications are also available. In general, red LEDs use less current for the same intensity than other colours.

LEDs make ideal indicators, running cool and being very robust. They can easily be driven directly off logic outputs. Figure 9.3 shows simple TTL circuits. Figure 9.3a lights the LED for a 1 input and fig. 9.3b lights the LED for a 0 input. The latter uses the current limiting characteristic of a TTL high level. The gate outputs in fig. 9.3 should not be used as inputs to other gates as the levels are degraded by the relatively high LED currents. It is usually convenient to use a 7404 or 7406 as an LED driver.

The LEDs of figs 9.3a and b are low intensity. Higher intensity can be obtained by using a transistor stage as in fig. 9.3c. Where several LEDs are to be driven it is usually economic to use a multi-transistor IC (e.g. 75491/75492) and a thin-film resistor package. A multi-indicator display using just four packages in addition to the LEDs is shown in fig. 9.4. Bar LED displays are particularly useful for multi-indicator applications.

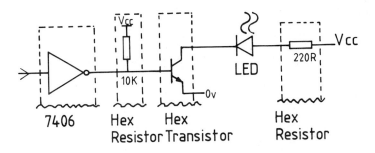

Fig. 9.4 LED driving with DIL packages.

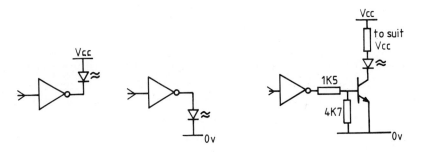

Fig. 9.5 Driving LEDs with CMOS. (a) On for input high. (b) On for input low. (c) Higher current.

In section 2.7 it was shown that a CMOS output looks like a resistor connected to the requisite supply rail. This resistor has a typical value of 1K ohm. An LED can be driven directly from a B series CMOS gate using the gate output impedance to define the current. The LED in fig. 9.5a lights for a 1 into the inverter and fig. 9.5b for a 0. The gate outputs should not be used as inputs to other gates because of the degraded levels.

On 5 volt supplies the LED is not very bright. If higher current is required or an A or UB (unbuffered) gate is used the circuit of fig. 9.5c is preferred. The circuit shown lights the LED for a 0 input. By using a pnp transistor with emitter to Vcc, the LED will light for a 1 input.

9.2.2 Liquid crystal displays (LCD)

Liquid crystal displays are particularly attractive for battery operated systems because of their minimal current requirements. LCD operation is complex, but is essentially based on materials which exhibit a regular crystal-like structure, even in a liquid state. The material is normally transparent, but when an external electric field is applied, complex interactions between molecules cause turbulence. The liquid then becomes opaque.

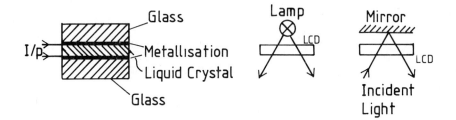

Fig. 9.6 The liquid crystal display. (a) Construction. (b) Transmissive mode. (c) Reflective mode.

An LCD is not a light emitter in the same way as other devices in this section are. Some form of external light source is needed, and this is normally obtained in two ways. In fig. 9.6b, known as transmissive mode, the LCD incorporates a light source which is viewed through the display. Energised segments appear dark. Figure 9.6c, called the reflective mode, is the commonest because it requires negligible current. The LCD is backed by a mirror, and ambient light used as illumination. Again, an energised segment appears dark.

In theory, LCDs are low voltage DC devices. In practice however, DC causes polarisation effects which quickly reduce the contrast of the device. To avoid polarisation the DC must be kept below 25 mV. AC drive is therefore used.

The commonest (and simplest) AC drive uses an oscillator and XOR gates as fig. 9.7a. The voltages for a display when turned off or on are shown on fig. 9.7b. The display has either 0 V across it, or AC with zero DC component. For fig. 9.7 to work correctly, the driving waveform must have a fairly even mark/space ratio. This is difficult to ensure with an oscillator and is usually obtained by a divide-by-two circuit using a D-type or JK flip-flop. The CMOS 4543 seven-segment decoder/driver incorporates the phasing circuits to drive LCDs directly.

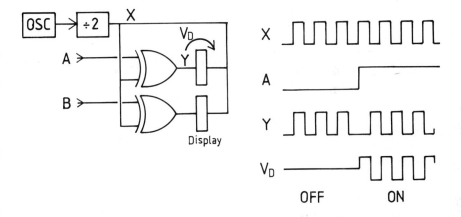

Fig. 9.7 Driving LCDs without DC bias. (a) Circuit. (b) Operation.

It is not possible to obtain a single LCD indicator (as it is to obtain a single LED). LCDs are usually found in multi-digit, seven-segment and other alphanumeric displays. Often these include the oscillator and XOR gates to produce the AC drive. The use of LCDs is described in sections 9.3 and 9.4 below.

9.2.3 Incandescent (filament) displays
Despite the obvious advantages of LEDs (low current, cool running, robust construction) the common filament lamp bulb is still used where high intensity or large physical size is required. The disadvantages of incandescent displays are inherently short life and their high current requirements. Only the lowest powered bulbs (e.g. 40 mA indicator lamps) can be driven directly off digital ICs. Usually a separate driver stage such as those shown in fig. 9.8 need to be used.

Filament displays exhibit an inrush current of up to ten times the operating current. This inrush current can last a few milliseconds as the display heats up. To avoid noise problems it is good practice to separate lamp currents and logic currents in circuits by using different power rails which only meet at one central point (a topic discussed further in section 13.3).

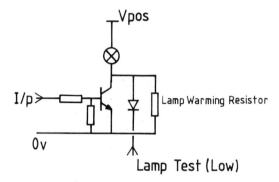

Fig. 9.8 Driving incandescent displays.

Inrush currents can be avoided by using lamp warming resistors as shown in fig. 9.8. The resistors should be chosen to give a lamp current between 0.1 and 0.25 times the operating current. Lamp warming resistors also increase a bulb's life, but increase the circuit's standing current.

Seven-segment filament displays are also available. These can be driven directly off digital ICs such as the 7447 (TTL) and 4511 (CMOS) seven-segment decoder/drivers. Failure of one segment is a problem with filament displays, and the test input on the decoders ICs should always be used.

9.2.4 Gas discharge displays

Gas discharge displays are used to form seven-segment and alpha-numeric displays. They consist of a cold cathode gas-filled indicator tube similar in principle to fluorescent lamps. The displays are bright orange and of very high intensity. They are ideal for viewing from a distance or in high ambient light.

The major disadvantage is the high supply voltage of 180 V DC. This supply has to be provided, and the output devices need to be capable of switching 180 volts. Buffer transistors are often required. The DS8880, (with TTL compatible inputs) and the DS8887/8889 (with CMOS compatible inputs) will interface directly with gas discharge displays.

The 180 volt supply needs only to provide a few milliamps. Encapsulated 5 volt to 180 volt DC-to-DC inverters are available at a reasonable price. These allow gas discharge displays to be used in systems with just a 5 volt supply. It should be noted that the 180 volt supply introduces a shock hazard into digital circuits.

9.2.5 Nixie tubes

The Nixie tube is a close relative of the gas discharge tubes. They have a mesh anode and shaped cathodes as shown in fig. 9.9. When a cathode is connected to 0 V, an orange discharge in the shape of the cathode is formed. The display is viewed through the mesh anode.

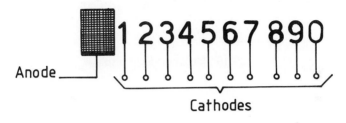

Anode

Cathodes

Fig. 9.9 The Nixie tube.

Nixie tubes give a more natural display than the somewhat artificial-looking seven-segment or dot displays. The cathodes are, however, at different distances from the tube front which causes odd parallax effects from some angles.

Nixies require a high voltage supply similar to gas discharge displays, and are driven in a similar way. The only difference is that a binary to decimal decoder is required. Nixie is a registered trade name of the Burroughs Corporation.

9.2.6 Phosphorescent displays

These displays are available in seven-segment form and are distinguished by a blue-green colour. They are rather similar to a thermionic valve with a heated cathode, a grid and several anodes coated with a fluorescent phosphorus. The selected anode is taken to a high voltage (typically 12-24 volts). As current flows, the phosphorus on the anode glows.

The grid is common to all segments and is held slightly negative to prevent leakage currents causing all segments to emit a slight glow in the off state. A seven-segment phosphorescent display can be represented by fig. 9.10.

Phosphorescent displays are aesthetically more pleasing than LED or LCDs, but are more complicated to drive. High anode switching to a supply in the range 15 to 24 volts is needed, with a negative grid bias supply and a cathode heater supply.

A phosphorescent display interface IC which can be driven by CMOS or TTL is the ICM 7235. This decodes four decades of BCD to seven-segment form.

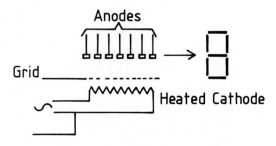

Fig. 9.10 Phosphorescent displays.

9.3 Alphanumeric displays

9.3.1 Seven-segment displays

A seven-segment display consists (oddly enough) of seven indicators arranged as in fig. 9.11. The indicators are usually LEDs or LCDs, but filament and cold cathode versions are also available. Large 'scoreboard' style displays can be built using bulb arrays or even fluorescent tubes. Regardless of type or size, the segments are coded a-f as in fig. 9.11.

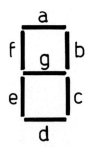

Fig. 9.11 Standard seven-segment labelling.

A seven-segment display can display the numerals 0-9 and the letters A-F for use in circuits using hexadecimal. The segments are illuminated as fig. 9.12a. Alternative arrangements for 6 and 9 are shown on fig. 9.12b. Although the latter are aesthetically more pleasing, the commonest figures are those in fig. 9.12a, because there is less ambiguity if segment failures occur.

Fig. 9.12 Seven-segment display formats. (a) Standard displays. (b) Variations. (c) Hexadecimal additions.

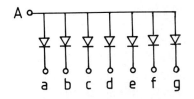
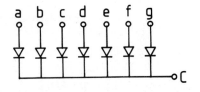

Fig. 9.13 Common anode/cathode displays. (a) Common anode. (b) Common cathode.

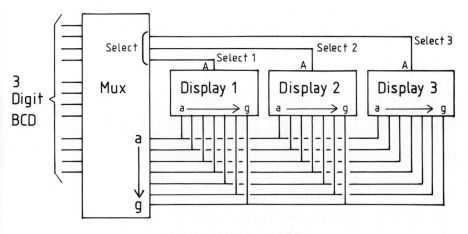

Fig. 9.14 Multiplexed LEDs.

LED seven-segment displays may be described as common anode (as fig. 9.13a) or common cathode (as fig. 9.13b). Common anodes are used where the LEDs are switched low, common cathode when they are switched high. Series resistors should be used with each segment.

Multi-digit LEDs commonly use one common set of input pins and digit select lines similar to fig. 9.14. A multiplexing arrangement, described in section 9.4, drives each digit in turn at a rate faster than the eye can follow.

Seven-segment displays are available in a range of sizes from 8 mm to 40 mm and up to eight digits.

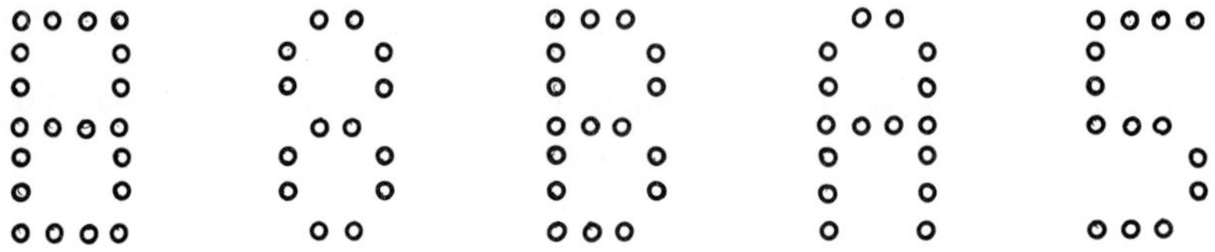

Fig. 9.15 Dot hexadecimal display.

A close relative of the seven-segment display uses dots as shown in fig. 9.15. This gives improved readability. At the time of writing, dot displays are only available with internal logic (i.e. the display accepts four-bit binary directly and contains its own decoder driver logic).

9.3.2 16-segment displays

A 16-segment display, shown in fig. 9.16, can display all alphabetical and numerical symbols plus some punctuation. Sixteen-segment displays are only available with internal logic, accepting seven-bit binary. Many incorporate internal storage and readback facilities.

Fig. 9.16 16-segment display.

9.3.3 Dot matrix displays

Full alphanumeric displays can also be obtained by a dot matrix similar to fig. 9.17. Common sizes are 5 × 7 and 7 × 9. Displays using dot matrices usually have internal logic and accept ASCII-coded binary, but displays can be constructed using discrete LEDs and dot matrix decoders such as the MCM 6576. These decoders are designed for VDU applications (see section 9.7) and only provide one row or column at a time. A quite complex multiplexing arrangement is therefore needed.

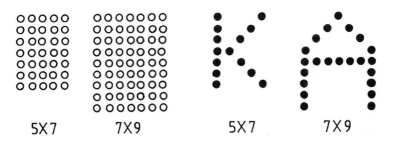

5X7   7X9   5X7   7X9

Fig. 9.17 Dot matrix displays. (a) Dot matrix. (b) Examples.

9.4 Display decoders and drivers

A decoder/driver is used to interface display devices to a logic system as shown in fig. 9.18. It is required to convert a binary input to the display requirements (e.g. BCD to seven-segment) and provide an output capable of supplying the display's current and voltage.

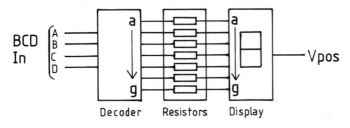

Fig. 9.18 Driving a seven-segment display.

Most decoder drivers are designed for use with seven-segment displays (more complex 16-segment and dot matrix displays usually use their own internal decoding logic). The commonest devices are the 7447 (and its derivations) in TTL; the 4511 (for LEDs) and 4543 (for LCDs) in CMOS. There is no direct LCD driver in TTL; the 4543 can, however, be driven off TTL. A gas discharge driver is not available in TTL or CMOS, but the 8880 decoder/driver can be driven off either family.

These devices differ in detail and drive capability and are described in section 9.8 and chapter 14. In essence, however, all are similar to fig. 9.19. The input data is presented as four binary digits ABCD (A least significant). The outputs drive the segments a-f as in fig. 9.11. Some decoders (the 7447 for example) have *low* outputs to turn a segment on. These require common anode LEDs. Devices with *high* outputs to turn a segment on (such as the 4511) require a common cathode LED.

The ripple blanking input/output connections are provided on most decoders and are used for the suppression of leading zeros (i.e. ensuring that, say, 96 on a four-digit display does not appear as 0096). If the ripple blanking input is low, the display outputs are blanked when a BCD input of zero is accepted. At the same time, if the display is blanked the ripple blanking output goes low. By linking blanking inputs and outputs as shown on fig. 9.20, leading zero suppression is obtained. If a display contains decimal points, trailing zero suppression (i.e. 7.200 appearing as 7.2) can be obtained by linking ripple blanking out to ripple blanking in on the next most significant stage.

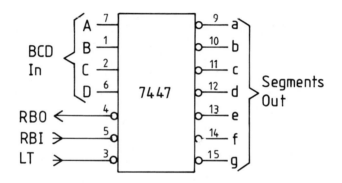

Fig. 9.19 The 7447 BCD to seven-segment decoder/driver.

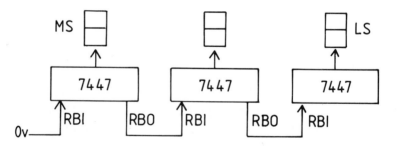

Fig. 9.20 Ripple blanking (leading zero suppression).

On the 7447 family, the ripple blanking output is an open collector. If an open collector gate is connected to the blanking *output*, a 0 will blank the display. This can be used to blank a display or vary the intensity. Other devices (such as the 4511) do not have ripple blanking, but have a single blanking input instead.

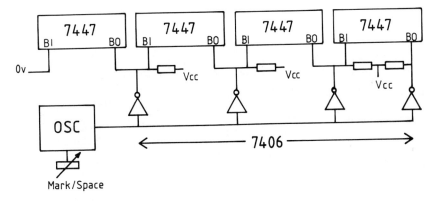

Fig. 9.21 Intensity control via blanking input/outputs.

If a blanking input (or RBO on the 7447 family) is driven from a constant frequency variable mark/space oscillator, the apparent intensity of the display will depend on the mark space ratio. Figure 9.21 shows a four-digit display using 7447s with ripple blanking and variable intensity. Intensity control cannot be used with LCDs.

Most decoders have a lamp test input. This simply turns on all seven segments (thereby displaying 8). If the lamp test input is not actually brought out to a panel push-button, it is a good idea to disable it via a resistor to the required voltage level, and provide a test pin so it can be enabled for test purposes with a jumper lead.

9.5 Multiplexing displays

Time-division multiplexing of displays can be used to reduce the number of ICs used and reduce the complexity of the decoder/display interconnections. It is particularly attractive where the display is remote from the rest of the circuit.

A typical multiplexed scheme is shown in fig. 9.22. This has four displays driven off four BCD inputs. The four inputs are applied to one of four data selectors IC1-IC4. IC1 takes all the A inputs, IC2 the B inputs, etc. An oscillator running at about 1 kHz drives a two-bit counter IC5. The counter outputs (two bits) go to the select lines of the data selectors and a decoder IC6.

The four decoder outputs select transistors TR1-TR4, which are connected to the anodes of four common anode LED seven-segment displays. The segment cathodes a-g are paralleled and driven off one decoder IC7. The decoder gets BCD data from the data selectors.

As the oscillator runs the counter selects A_0, B_0, C_0, D_0 then

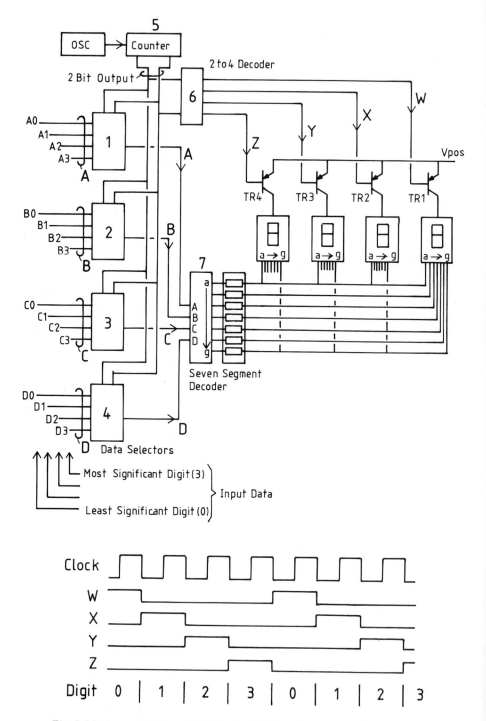

Fig. 9.22 Complete four-digit display. (a) Circuit diagram. (b) Waveforms.

A_1, B_1, C_1, D_1 and so on, cycling through all four BCD inputs. The seven-segment decoder provides the segment data for each decade in turn. At the time when a decade's data is on the common segment lines, the corresponding transistor TR1-TR4 is turned on by the decoder IC6.

The displays are thus being strobed rapidly, and at any time only one display is on. The eye cannot follow this, however, and all four displays appear to be on together, displaying the correct data.

Note that four multiplexed displays require just eleven connections from the circuit; seven segments and four digit select lines. Each additional display requires just an additional digit select line. Pin limitation means that most multi-digit displays use multiplexing.

Multiplexing can be easily used with LED, gas discharge, phosphorescent and incandescent displays (the latter require diodes in each display lead to prevent sneak paths). Multiplexing of LCDs is possible, but more complex due to the need to provide an AC drive and the slow display response. Usually multi-digit LCDs have their own internal decoders and the BCD data is multiplexed to the display.

Multiplexing does have some disadvantages. Because each digit is only on for a fraction of the time, brightness is noticeably reduced. This can be overcome by overdriving the display; increasing the voltage or increasing the current (dependent on the type). Overdriving can decrease the display life, however. Occasionally, in poor ambient light, beat flicker between multiplexed displays driven off different clocks becomes apparent. This can be overcome by using one common oscillator for all displays.

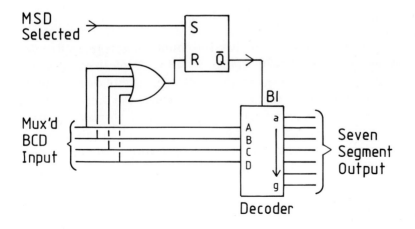

Fig. 9.23 Leading zero suppression on multiplexed displays.

Leading zero suppression can be added to multiplexed displays, with a few additional ICs. The principle is shown in fig. 9.23. The digits are scanned MSD to LSD. A flip-flop is set each time the counter selects the MSD, and is reset when a non-zero BCD digit is detected. The flip-flop output drives the blanking input of the seven-segment display. The display is thus blanked from the MSD until a non-zero digit is reached.

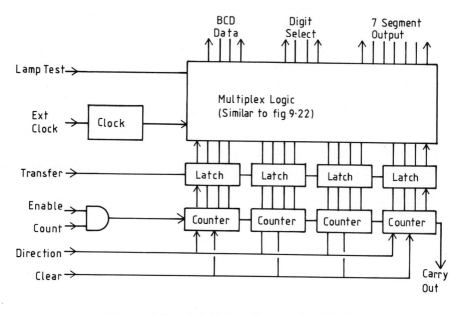

Fig. 9.24 Simplified block diagram of ZN1040.

9.6 Composite devices

There are many composite ICs for counting/storage/display driving that are not specifically CMOS or TTL. Typical of these is the Ferranti ZN1040 shown on fig. 9.24. This useful device contains a four-digit up/down counter, latches (which can freeze the display whilst counting continues), multiplexer and display drivers. The multiplexer oscillator is included in the IC, and an intensity control pin is provided. The device can be used with common cathode or common anode displays. A multiplexed BCD output is also available.

A useful LCD driver IC is the 7211 which accepts multiplexed BCD data (from, say, a ZN1040) and contains the necessary latches and drivers to drive directly a four-digit LCD display. An on-chip oscillator to give the necessary AC drive is provided.

Composite counter/displays are also available, where an alpha-numeric display, counter and latches are all included in the one device. A common device, with counter, latch and display, is the seven-segment TIL 306. A BCD output is provided for use in, say, batch counting applications.

There are many composite devices, and it would be pointless to describe them all. The more common devices are listed in section 9.8.

9.7 VDU fundamentals

A visual display unit, or VDU, is a commonly used display device. It is an interesting example of logic design, and demonstrates many useful techniques.

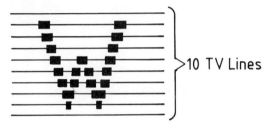

Fig. 9.25 One character on a VDU display.

A VDU produces a video output for a normal CCTV monitor. The 'picture' is composed of 625 lines which are 'drawn' 50 times a second (British standard, somewhat simplified). An enlarged portion of a screen with one character would appear as in fig. 9.25.

Let us assume a VDU display of 72 X 24 characters (i.e. 24 rows of 72 characters per row). These characters are stored in a memory of 24 X 72 X 8 bits. It is convenient to consider this store as being addressed in the same way as the characters appear in the display. A location is addressed by an X address, which gives the character position in a row, and a Y address which gives the row.

Characters are displayed on a dot matrix, common arrangements being 7 X 10, 7 X 9, and 7 X 5. For simplicity we shall assume that one line of the TV display corresponds to one line of the dot matrix and a 7 by 9 display.

A character row will be displayed one line at a time, so the TV signal will consist of:

Row 1 line 1 character 1 . . . line 1 character 72

    :   :

    line 9 character 1 . . . line 9 character 72

and so on for all 24 rows.

The characters in the VDU store are converted to their dot patterns by a ROM called a character generator. It might be thought that this would have an eight-bit input and 63 outputs (for the 7 X 9 elements of the dot matrix).

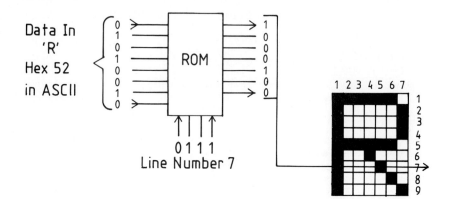

Fig. 9.26 Character generator ROM.

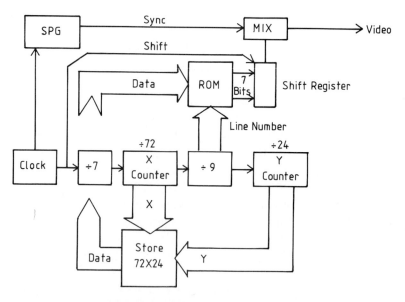

Fig. 9.27 VDU block diagram.

All that is required, however, is the seven outputs for one line at a time, so the character generator has an eight-bit data input and a four-bit line number. The seven outputs give the bit pattern for one line as shown in fig. 9.26. The character generator output is loaded to a shift register, and shifted out one bit at a time to form the serial TV signals.

The complete VDU block diagram is shown in fig. 9.27. The sequence starts with a clock running at the bit rate of the dot matrix. This clocks the shift register and a divide-by-seven counter. Since a character contains seven dots per line, the X address counter is stepped every seven dots and can be used as the X store address.

Each character occupies nine lines, so the X counter is followed by a divide-by-nine counter which provides the line number for the character generator ROM. This is followed by the Y counter which gives the Y address for the store. The timing chain brings characters from the store to the character generator at the correct point in time to position the character on the required place on the screen.

The bit pattern corresponding to the serialised dot matrix is mixed with sync pulses from the sync pulse generator. These sync pulses instruct the TV monitor when to start a new line and new picture. The resulting signal, called composite video, can now be fed direct to a CCTV monitor.

The above description is somewhat simplified, but all VDUs (even complex ones with colour graphics) utilise the principle of fig. 9.27 with its store, timing chain and character generator.

9.8 Display-related ICs

9.8.1 TTL

Decoders (denotes open collector output)*

Number	Description	Type	Voltage	Current mA
7442	BCD to decimal	N, L, LS	Logic Levels	
7443	XS3 to decimal	N	Logic Levels	
7444	XS3 Gray to decimal	N	Logic Levels	
7445*	BCD to decimal	N	30	80
7446*	BCD to seven-segment	N	30	80
	—	L	30	20
7447*	BCD to seven-segment	N	15	40
	—	L	15	20
	—	LS	15	24
7448*	BCD to seven-segment true logic output	N, LS	Logic Levels	
7449*	BCD to seven-segment	LS	5.5	8

Number	Description	Type	Voltage	Current mA
74141*	BCD to decimal cold cathode tube	N	60	7
74145*	BCD to decimal	N	15	80
–		LS	15	80
74246*	BCD to seven-segment	N	30	40
74247*	BCD to seven-segment	N	15	40
–		LS	15	24
74248*	BCD to seven-segment	N	Logic	Levels
74249*	BCD to seven-segment	N	5.5	10
–		LS	5.5	8
74347*	BCD to seven-segment	LS	7	24
74445*	BCD to seven-segment	LS	7	24
74447*	BCD to seven-segment	LS	7	24

Counters with display facility

Number	Description	Type	Clock	Voltage	Current mA
74142	BCD counter/latch, Nixie driver	N	20↑	60	7
74143	BCD counter, seven-segment driver	N	12↑	7	15
74144	BCD counter, seven-segment driver	N	12↑	15	25

74143 BCD counter, seven-segment driver, has constant current output, no series resistor needed.

9.8.2 CMOS
Decoders

Number	Description	Voltage	Current mA
4055	BCD to seven-segment LCD with separate display supply	15	2
4056	BCD to seven-segment input latch, separate display supply	15	2
4511	BCD to seven-segment active high	Vdd	5
4513	BCD to seven-segment with ripple blanking	Vdd	5
4543	BCD to seven-segment LED driver	–	–

Counters with display facility

Number	Description	Clock Mhz	Voltage	Current mA
4026	Decade counter — seven segment output	2.5↑	Vdd	0.5
4033	Decade counter — seven-segment output	2.5↑	Vdd	0.5

CHAPTER 10

Event Driven Logic

10.1 Introduction

Many logic systems are driven by randomly occurring external events. Typical of these are industrial control systems using inputs from limit switches, push-buttons and providing on/off outputs to relays, hydraulic valves and similar devices. Similar circuits occur in computing where the control circuits have to handle random signals from peripherals, and the memory. Such systems are called event driven logic. This chapter outlines design procedures for event driven logic systems.

10.2 State diagrams

An event driven logic system can be represented by fig. 10.1. Superficially this is identical to the combinational logic diagram of fig. 3.1a, but there is one very important difference. In a combinational logic circuit the output states are defined solely in terms of the inputs, and a given input combination always produces the same outputs.

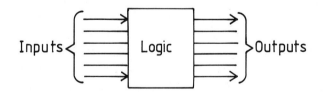

Fig. 10.1 Representation of event driven logic.

In event driven logic the output states depend on both the input states and the previous state of the circuit. A given input combination may not always produce the same outputs.

In section 4.1 a simple motor starter circuit was described. This had two inputs, a start button and a stop button, and a single output to the motor contactor. Obviously the start button starts the motor and the stop button stops the motor. If the start button and stop button occur together the motor stops.

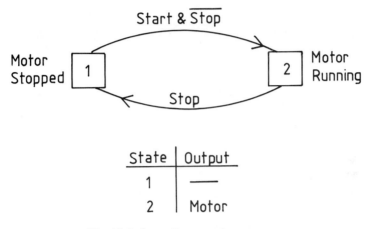

Fig. 10.2 State diagram of motor starter.

This operation cannot be achieved with combinational logic because the input state with neither button pressed can result in either the motor running or the motor stopped according to which button was pressed last. The required circuit is best described by a state diagram as in fig. 10.2.

There are two states that the circuit can be in; motor stopped and motor running, which we can call state 1 and state 2. To go from state 1 to state 2 we must have the input combination 'Start and Not Stop'. To go from state 2 to state 1 we require the input 'Stop'. Associated with the state diagram we have an output table which shows that the single output is energised in state 2. Figure 10.2 is a complete description of the operation we require.

A motor starter is a very trivial example, which can be realised by a single flip-flop without the need for detailed analysis, but the principle of fig. 10.2 is the basis of more complex circuits.

State diagrams were introduced in section 7.4.3 when non-binary counters were discussed. There are some important differences between event driven state diagrams and counter state diagrams. A counter state diagram such as the decade counter of fig. 10.3 only responds to one input; an event driven state diagram responds to many inputs.

Each state of a counter diagram only leads to one next state. On the decade counter state diagram of fig. 10.3, for example, state 7 leads only to state 8 (bidirectional counters are a special case where there are two possible next states). A given state of an event driven circuit can lead to several next states according to the input arriving at the circuit.

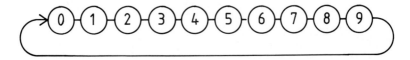

Fig. 10.3 State diagram for decade counter.

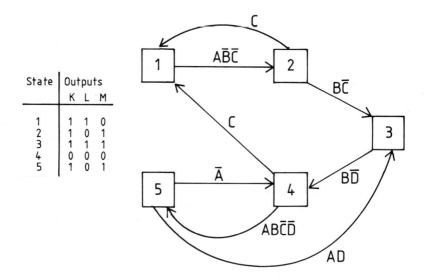

State	Outputs		
	K	L	M
1	1	1	0
2	1	0	1
3	1	1	1
4	0	0	0
5	1	0	1

Fig. 10.4 Multi-route state diagram.

On fig. 10.4, for example, state 4 can lead to state 5, if the input combination A = 1, B = 1, C = 0, D = 0 appears, or to state 1 if input C = 1 appears. Other input combinations will cause the circuit to stay in state 4.

Care must be taken to avoid ambiguities in laying out a state table. In fig. 10.5a, for example, state 3 can go to state 1 if A = 1 or to state 4 if C = 1. The transition from state 2 to 3 does not depend on A or C, so the circuit could arrive in state 3 with A = 1 and C = 1. As drawn, it is not clear whether state 4 or 1 would follow state 3. Exit conditions from a state should be mutually exclusive.

Care should also be taken to ensure that a circuit cannot oscillate between states. In fig. 10.5b the circuit can oscillate between states 1 and 2 if A = 1, B = 1, C = 0. It could be that the designer really intended the transition from state 1 to 2 to be A = 1, B = 1, C = 1.

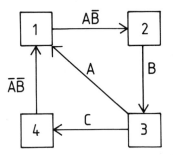
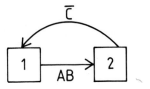

Fig. 10.5 Incorrect state diagrams. (a) Ambiguous state diagram. (b) Potentially oscillatory state diagram.

Often these requirements can be ignored if certain combinations cannot physically occur. Figure 10.5a could be a legitimate state diagram if it was physically impossible for A = 1 and C = 1 to occur together. (They could, for example, be two signals saying 'Disc Input Selected' and 'Magnetic Tape Input Selected'.)

On fig. 10.5a, the transition from state 2 to state 3 does not depend on A. If the circuit arrives at state 3 with A = 1 it will immediately go onto state 1, glitching any outputs from state 3 as it passes. Whether this is a problem or not depends on the application, but it can be avoided by careful layout of the state diagram.

Having drawn up a state diagram and output table for the required operation, the next stage is to produce a working circuit. There are several ways in which this can be achieved.

10.3 From state diagram to circuit

10.3.1 Non-optimal solutions
The obvious way to translate a state diagram to a circuit is to allocate one flip-flop to each state. Simple combinational circuits are used with the inputs to set and reset the flip-flop. With fig. 10.4, for example, state 1 flip-flop would be:
(a) Set by (C AND state 2) OR (C AND state 4).
(b) Reset by (A AND $\bar{B}$ AND $\bar{C}$) i.e. state 2.
Similar combinations are constructed for each state. There will be one combinational circuit for each line on the state diagram. The full circuit corresponding to fig. 10.4 is shown in fig. 10.6.

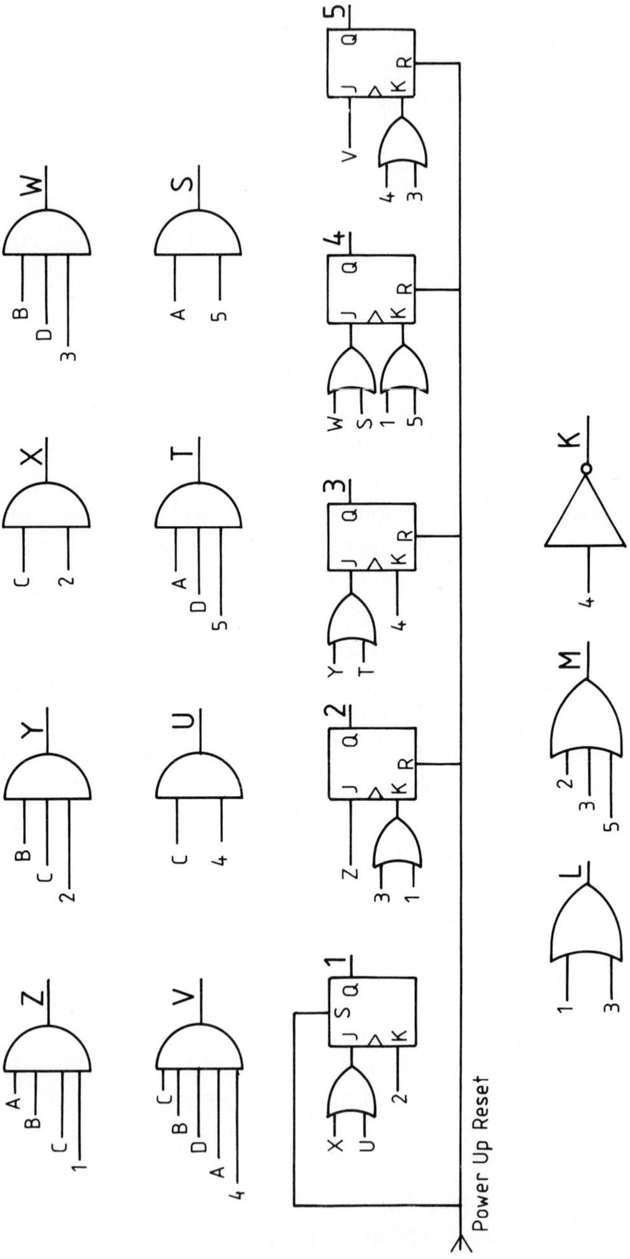

Fig. 10.6 Circuit corresponding to fig. 10.4.

Simple RS flip-flops probably would work, but potential race conditions exist. The signal setting state 2, for example, is also clearing state 1. Which occurs first is a matter of relative propagation delays. If state 1 clears before state 2 sets, the circuit could end up in no state. It is obviously preferable to use edge triggered synchronous RS or JK flip-flops, which will not suffer from race conditions. If the rest of the circuit is not synchronous and a clock is not already available, all that is required is a simple oscillator constructed from, say, a 555.

Outputs are obtained by ORing the requisite stages as shown. Note that it is simpler to produce K by $\overline{4}$ rather than (1 or 2 or 3 or 5).

This design method has some merit; it is simple to design, easy to understand and easy to modify. It does, however, use more devices than are strictly necessary; a more optimal solution would use just three flip-flops.

The technique also has some potential hazards. The five-state flip-flops can take up 32 different combinations, only five of which are used. At power up, or via noise, the circuit could take up any of the 27 unused combinations. Even worse, many of these invalid combinations will cause the circuit to be in two states simultaneously. Some form of power up reset or direct reset should be provided. Figure 10.6, for example, uses a direct reset to state 1.

10.3.2 Solutions using secondary variables.

Figure 10.4 has five states, and using the design technique in the previous section a solution using five flip-flops was obtained. Five states can be defined by three flip-flops, so it should be possible to design a suitable circuit with three flip-flops. The three outputs from these flip-flops are called secondary variables. We shall denote them Z, Y, X and allocate them as follows:

State	Z	Y	X
1	0	0	1
2	0	1	0
3	0	1	1
4	1	0	0
5	1	0	1

This allocation is purely arbitrary. For each particular problem there is an 'optimal' allocation, but it usually is not worth searching for it. Note that there are three unused combinations of the secondary variables in this particular problem.

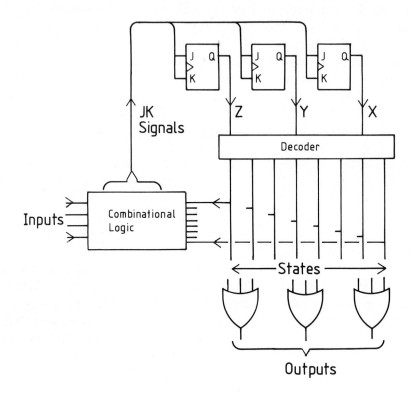

Fig. 10.7 Block diagram of solution using secondary variables.

A block diagram of the solution is shown in fig. 10.7. A synchronous JK circuit is used again, to avoid races. The secondary variables are decoded to give eight outputs (only three of which are used). Combinational logic is used on XYZ or the decoder outputs, along with external inputs to produce the JK inputs for the three flip-flops.

Figure 10.4 is redrawn in fig. 10.8 with XYZ added to the state diagram. By inspection we can see that:

X is set on transitions $2{\to}1$, $2{\to}3$, $4{\to}1$, $4{\to}5$
X is reset on transitions $1{\to}2$, $3{\to}4$, $5{\to}4$
Y is set on transitions $1{\to}2$, $5{\to}3$
Y is reset on transitions $3{\to}4$, $2{\to}1$
Z is set on transition $3{\to}4$
Z is reset on transitions $5{\to}3$, $4{\to}1$

Eight combinational circuits are required (one for each used transition), which are combined as before to give the required J and K inputs. Outputs are simply obtained by an OR combination of the decoded secondary variables.

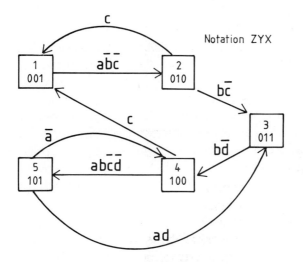

Fig. 10.8 Secondary variable version of fig. 10.4.

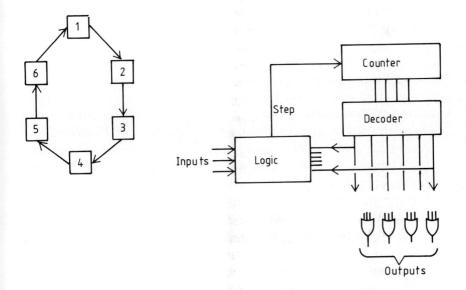

Fig. 10.9 Single route state diagram. (a) Ring state diagram. (b) Block diagram.

A secondary variable solution is less straightforward to design than the previous technique, but uses fewer ICs. The difference was marginal for our five-state problem, but becomes more significant with more states. The technique can still have unused combinations of secondary variables, however, and care should be taken to ensure that the circuit cannot lock up at switch on.

10.3.3 Single route logic

Single route event driven logic has no alternative routes, and simply has a ring state diagram similar to the six-state of fig. 10.9a. Although these can be designed using either of the previous two methods, solutions using a simple counter are often preferred.

Figure 10.9b uses a binary counter and decoder to define the states, although any counter (even a shift register Johnson counter) could be used. The decoder outputs and the inputs necessary to move to the following states are gated with a free-running clock. When a state's transition conditions are met, the counter advances to the next state.

10.3.4 Time sequenced logic

Time frequently plays an important part in event driven logic. A sequence may be required to wait on a state for a given time, or exit a state if a certain input does not occur.

A typical example is shown in fig. 10.10, which is a state diagram for a gas burner controller. When the start button is pressed, a 15-second air purge is given. The pilot valve is opened and the igniter started for four seconds. If, at the end of this time, the flame detector shows the pilot to be lit, the main gas valve is opened. At any time the stop button terminates the sequence.

A non-valid signal from the flame detector at any time puts the system to an alarm state, which is exited by a reset push-button.

There are two timed operations in this sequence; the 15-second air purge and the four-second ignition/subsequent flame test. These can be implemented by using an output from the states to fire an edge triggered timer whose Q output is brought back as an input. If we denote $Q1$ as the output of the 15 second air purge timer, the transition from state 2 to state 3 becomes $F.Q1$. Similarly, if $Q2$ is the four-second ignition timer, the transition from state 3 to state 4 is $\overline{Q2}.F$ and $\overline{Q2}.\overline{F}$ from state 3 to state 5.

There are, in general, three possible timed operations:
(a) Wait in state N for time T (fig. 10.11a).
(b) After time T go to state P if input X present; otherwise go to

state Q (fig. 10.11b).

(c) When input X present go to state P; if input X does not occur within time T go to state Q (fig. 10.11c).

Each circuit assumes a timer triggered on entering state N, with an input T from the timer used as a transition condition. More complex timed operations are variations on (or combinations of) fig. 10.11.

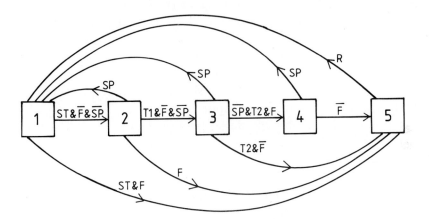

Inputs: Start (ST), Stop (SP), Flame (F), Reset (R), Timers (T1, T2)

State	Description	AIR	PILOT	IGN	GAS	TIMER1	TIMER2	ALARM
1	OFF	0	0	0	0	0	0	0
2	AIR PURGE	1	0	0	0	1	0	0
3	IGNITION	1	1	1	0	0	1	0
4	ON	1	1	0	1	0	0	0
5	ALARM	0	0	0	0	0	0	1

Fig. 10.10 State diagram and output table for gas burner.

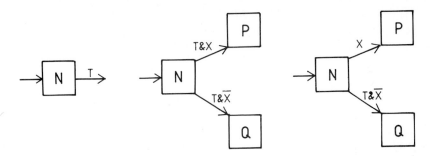

Fig. 10.11 Time-related events. (a) Wait for T. (b) Test after T. (c) Test for X.

10.4 General observations

Event driven logic uses flip-flops, timers and counters and as such is particularly vulnerable to noise. All the practical precautions given in chapter 13 should be followed.

Faults in event driven logic can be difficult to trace, and the designer should make every endeavour to assist the user. All inputs and outputs should have monitoring LEDs, and the condition of the state flip-flops should be monitored. A state diagram and output table should be part of the documentation.

Safety should be carefully considered when event driven logic is used to control potentially hazardous equipment. The circuit's behaviour under fault conditions should be analysed. Input and transducer failures can cause input combinations to appear which were not considered by the designer, and if due care has not been taken hazardous outputs could occur. Figure 10.10, for example, checks the flame detector at each stage and does not assume it is correct when the start button is pressed. True fail-safe design relies on multi-input majority voting circuits, a rather specialised subject.

CHAPTER 11

Communications and Highways

11.1 Introduction

Previous chapters have tacitly assumed that the various parts of a digital system are 'close' to each other, and problems associated with distance (noise or time delays, for example) can be ignored. In this chapter, the problems of operating digital systems in applications where parts are not 'close' are examined. The treatment will not be mathematically rigorous, being more concerned with practicalities than strict theory.

11.2 Transmission lines

Consider the circuit of fig. 11.1. We have some voltage source, a switch, a transmission line (e.g. coaxial cable or twisted pair) and a load represented by a resistor. To study transmission lines we need to be able to predict the currents and voltages flowing in the circuit when the switch is closed and opened.

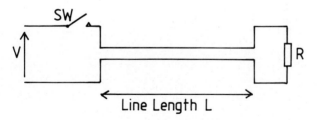

Fig. 11.1 Simple transmission line.

At the instant the switch is closed, because electrical signals travel no faster than the speed of light, the current flowing into the line is *not* necessarily given by I = V/R. It can be shown that a transmission line has a 'characteristic impedance', which can be considered to be the initial resistance seen by a signal entering a line. Domestic TV coax, for example, has a characteristic impedance of 75 ohms.

When the switch SW is closed, an initial current will flow into the line. The current will be I = V/Z where Z is the characteristic impedance. Note that this is totally independent of R. A current step of magnitude I and a voltage step V passes down the line at a speed approaching that of light. After a short time, the signal reaches the load. There is now a bit of an anomaly. The voltage step is V, the current step is I, but the current required by the load is V/R. If the value of the load is *not* the same as the characteristic impedance, a partial reflection of the signal will occur. We can define a 'reflection coefficient', ρ, as:

$$\rho = \frac{\text{reflected voltage amplitude}}{\text{incident voltage amplitude}}$$

It can be shown that for a characteristic impedance Z and load R

$$\rho = \frac{R - Z}{R + Z}$$

for R = 0 (short circuit) ρ = −1
for R = ∞ (open circuit) ρ = +1
and for R = Z ρ = 0, no reflection

The magnitude of the received voltage step is *initially* $V(1 + \rho)$. If a reflection occurs, however, further reflections will occur at *each* end of the line (the apparent load of a voltage source is a short circuit). Each reflection will be of smaller magnitude than its predecessor, until Vo = V and I = V/R.

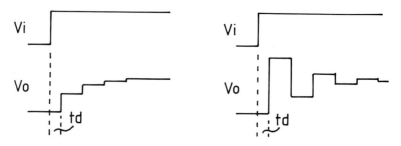

Fig. 11.2 Waveforms for incorrectly terminated lines. (a) R < Z. (b) R > Z.

Possible waveforms for R < Z and R > Z are shown in fig. 11.2. Both have implications for digital systems if they are connected to gate inputs. Figure 11.2a spends some time in the grey area between

0 and 1 states and would be very vulnerable to noise. Figure 11.2b actually has several transitions between 0 and 1 states which could cause chaos in any circuit using counters or clocks.

The above ideas suggest that all transmission lines should be terminated with the characteristic impedance of the line. All conductors, whether coaxial, PCB tracks, single wires or whatever, have a characteristic impedance, so this would appear to be a major restriction. There are some mediating factors, however.

Examination of fig. 11.2 shows that the duration of each reflection is equal to twice the transition delay experienced by the signal whilst passing down the line. Gates, in general, do not respond to signals that are shorter than the gate propagation delay so it can be assumed that reflections can be ignored if the gate propagation delay is longer than twice the line transition delay.

The second mediating factor arises from the finite rise and fall times of real signals compared with the instantaneous steps of fig. 11.2. If the signal rise time is longer than the line transition time, the magnitude of the reflections will be reduced (in reality, not just as far as the receiving gate is concerned).

The consideration of a conductor as a transmission line therefore depends on the rise time of the signal and the propagation delay of the receiving device. In general, a line should be terminated if:

t > one-quarter of propagation delay of receiving gate
or t > one-quarter of signal rise time
where t is the signal transition time down the line.

CMOS, for example, has a propagation delay of about 100 ns and rise and fall times of about 40 ns (both will, of course, rise with load capacitance). Signals between CMOS gates need not be terminated for line transition times of less than 25 ns. Assuming that the signal travels down the line at the speed of light, 3×10^8 ms^{-1}, we can find the corresponding line length

$$L = v \times t$$
$$= 3 \times 10^8 \times 25 \times 10^{-9}$$
$$= 8 \text{ m}$$

CMOS can be used up to about 8 m without worrying about transmission line effect (although other noise effects could reduce this further).

Standard TTL has propagation delays of about 10 ns and rise and fall times of around 5 ns (faster for Schottky). Using similar procedures we obtain a safe maximum length of line of about 0.8 m.

ECL has propagation delays as low as 1 ns and similar edge speeds. High speed ECL requires transmission line techniques for lines as short as 8 cm. ECL consequently does not respond well to haphazard board layouts.

It should be emphasised that the above considerations apply to transmission line effects, and in practical systems external noise may serve to make matters worse. In general:

(a) CMOS can be used freely within a cubicle, but connections between cubicles should be dealt with as transmission lines.

(b) TTL in standard, LS and L form can be freely used on boards or within standard racks but may need transmission line treatment within cubicles.

(c) S or H TTL may be freely used within boards but may need transmission line treatment within racks.

(d) ECL board layout is critical, in high speed versions at least.

Where a transmission line approach is needed, it is necessary to establish a predictable and consistent characteristic impedance. Changes in characteristic impedance along a line will cause reflections. A single wire run haphazardly in space has an unpredictable characteristic impedance. Preferred lines are:

(a) Coaxial cable

(b) Signal lines (PCB track or wire) above a ground plane

(c) Ribbon cable with alternate signal/ground cores

(d) Twisted pairs

Characteristic impedances are normally in the range 50-600 ohms.

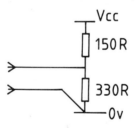

Fig. 11.3 Termination for TTL signals.

Tap-offs are allowed along a transmission line, but these should be of high impedance to minimise reflection. The line should, of course, be terminated correctly. A suitable termination for TTL levels and a 100 ohm line is shown in fig. 11.3. This requires a driver that can sink 30 mA, which is within the capabilities of some open collector drivers such as the 7406. Alternatively, discrete driven or special line drivers (described in the next section) could be used. Spurs off a transmission line must be kept short or half the characteristic impedance will be seen at the junction and reflections will result.

An alternative method of termination, called 'series termination' can be applied at the transmitting end. The equations developed above for transmission lines show that the initial step received at an open circuit termination is twice the input step. On fig. 11.4, we have a series resistor R at the transmitter, equal to the characteristic impedance of the line. The initial step, Va, is half Vin, but the output Vb equals Vin. The reflection from the open circuit is absorbed by R (which correctly terminates the line) giving the waveforms shown on fig. 11.4b.

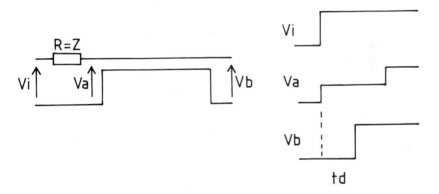

Fig. 11.4 Series termination. (a) Circuit arrangement. (b) Waveforms.

Series termination requires a high input impedance receiver, but uses considerably less power than the termination arrangement of fig. 11.3. There can, however, be no tap-offs along the line as the signal is severely degraded at points other than Vin and Vb.

11.3 Line driver ICs

When long lines are driven, correct termination is not the only problem. In most systems, long lines will be vulnerable to noise induced both from adjacent signals and external sources such as power cables, motors, transformers, etc.

A somewhat simplified picture of induced noise is shown in fig. 11.5a. A signal is transmitted down a pair of lines from left to right. The top line is at potential V2 to ground and the bottom line at V1. The signal has a voltage (V2 − V1). As the signal passes down the line, noise Vn is introduced. Because the lines are in close proximity, preferably twisted together, the same noise is introduced in each line.

The received voltages are (V2 + Vn) and (V1 + Vn) but the difference is still V2 − V1.

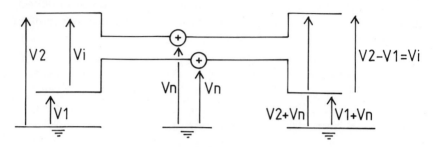

Fig. 11.5 Common mode noise.

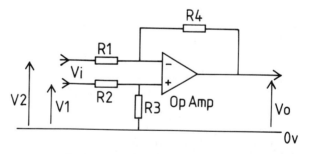

Fig. 11.6 Noise induced via 0 V lines.

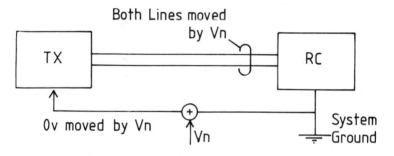

Fig. 11.7 Differential amplifier.

The noise in fig. 11.5 is known as common mode noise, and is the type of noise usually encountered. A similar effect can be caused by noise induced in 0 V lines as shown in fig. 11.6. Series mode noise occurs where currents are induced around the loop consisting of the transmitter lines and receiver. Series mode noise is difficult to remove electrically and is best dealt with by filtering at the receiving end or by screening.

Common mode noise, can, however, be almost completely removed by a device called a differential amplifier and shown in fig. 11.7. Although the theory is based on an analog device, it is equally applicable to digital signals. If the circuit is arranged with R1 = R2 and R 3 = R4 it can be shown that:

$$Vo = -\frac{R4\,(V2 - V1)}{R1}$$

In particular, if all resistors are equal Vo = −(V2 − V1), which is the condition required to remove common mode noise on an incoming signal. In practice, the gain is set to be about 0.1 and the differential amplifier is followed by a simple amplifier of gain 10. This arrangement allows the circuit to deal with a higher range of common mode noise.

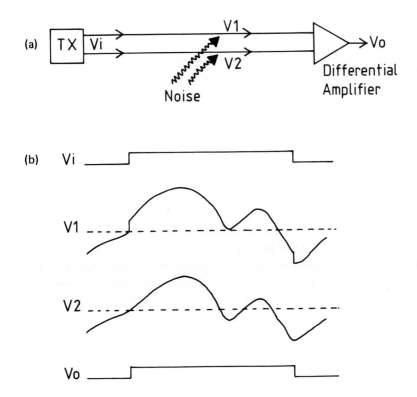

Fig. 11.8 Operation of a differential receiver. (a) Circuit diagram. (b) Waveforms.

A typical arrangement is shown in fig. 11.8a. A signal is sent down coaxial cable and is degraded by noise. At the receiving end, a differential amplifier is used to remove the noise. The differential amplifier should present a terminating resistor to the line.

Although it would be possible to use amplifiers built from, say, 741 or 709 Op Amps they would be rather slow. Devices called line drivers and line receivers fulfil the same purpose and are designed specifically for digital systems.

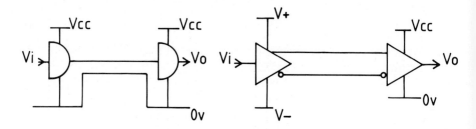

Fig. 11.9 Types of line driver/receiver. (a) Single-ended circuit. (b) Balanced circuit.

There are two basic types of line driver, shown in fig. 11.9. A single-ended circuit uses one signal line and a common ground return. A balanced circuit uses two signal lines of equal (but opposite) polarity (e.g. for a 1, line 1 is +15 V, line 2 is −5 V; for a 0, line 1 is −5 V, line 2 is +5 V). Balanced circuits give better noise immunity.

Neither of the common logic families have line drivers/receivers as such (although there are the 74128 and 74S140 which are single-ended line drivers, and some buffers have sufficient current sourcing/sinking capability to be used as line drivers. Texas Instruments have the 55107 line drivers/receivers in their LINEAR family). True line driver/receivers are available from a variety of suppliers. Typical is the Fairchild 75110 (line driver) and 75109 (line driver). These drivers are designed for use on balanced lines, and the 75110 has a tristate output allowing a bidirectional (half duplex) transmission line to be constructed. Note that *both* ends of the line are terminated on bidirectional lines. The receivers have a very high input impedance, so tap-offs along a line are possible without causing reflections.

11.4 Highway-based systems and buses

Computer and other 'number crunching' digital systems move numbers round internally on a parallel highway often called a bus (for busbar). Typically these can have 8, 16, or 32 signal lines. A typical microcomputer shown in fig. 11.10 has an eight-bit bidirectional data bus to carry data between the microprocessor, store and input/output; a 16-bit single direction address bus (to identify store or input output addresses); and a 12-bit mixed control bus (carrying timing signals, commands such as Read/Write and status signals). The concept of a highway allows an expandable system to be built; additional store or input/output can simply be 'bolted onto' the bus. Although originally evolved in computers, highways are found in a variety of equipment which is naturally 'modular' in form.

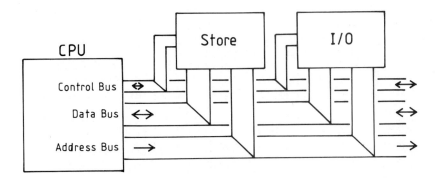

Fig. 11.10 Block diagram of bus-based microcomputer.

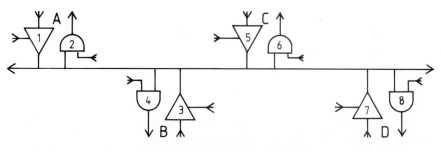

Fig. 11.11 Bus communications.

Highway systems are made possible by tristate gates (early highways used open collector gates which were wire-ORd onto the highway). One line of a highway could be represented by fig. 11.11. Gates 1, 3, 5, 7 are tristate buffers. Gates 2, 4, 6, 8 are straightforward AND gates. To pass data from location A to location C, say, tristate gate 1 is enabled and the select line on gate 6 taken to a 1. Data now passes from A to C. Communication between any two points on fig. 11.11 could be established in a similar manner.

Tristate gates and buffers are readily available in both TTL and

CMOS (74126 quad tristate buffer, 40244 octal tristate buffers). Devices called transceivers (such as the 40245 octal transceiver) employ back-to-back tristate gates as in fig. 11.12 and allow linking of two separate highways or provide increased fan-out.

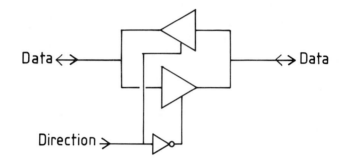

Fig. 11.12 Back-to-back (bidirectional) tristate gate.

Storage systems are often required to place data onto highway. To reduce the number of ICs in a circuit, both TTL and CMOS provide latches, D-types and JK flip-flops with tristate outputs (for example, 40373 octal transparent latch with tristate output, 74LS374 octal D-type with tristate output).

11.5 Data communications

11.5.1 Introduction
Communication between digital systems and devices such as VDUs, printers, etc. (and between digital systems themselves) employs techniques that are an extension of the ideas outlined above. This section examines the principles and standards used in data communications. Apart from the obvious applications in linking computers and peripheral devices, data communications are widely used in telemetry, television and remote control.

Data is normally transmitted in eight-bit 'words' and can be achieved in two ways, illustrated in fig. 11.13. Parallel transmission uses one line per bit plus timing signals. Because all bits are transmitted simultaneously, one word will be sent for each clock pulse. Serial transmission uses one line, and each bit of data is sent sequentially. Synchronising start and stop bits are necessary to ensure the transmitter and receiver work together. Usually ten or eleven clock pulses are required to send an eight-bit data word, so serial transmission is considerably slower (but cheaper) than parallel transmission.

It is particularly well suited to passing data over telephone lines or radio links, where parallel paths cannot be used.

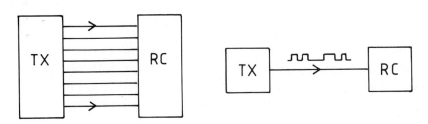

Fig. 11.13 Serial and parallel data transmission. (a) Parallel transmission. (b) Serial transmission.

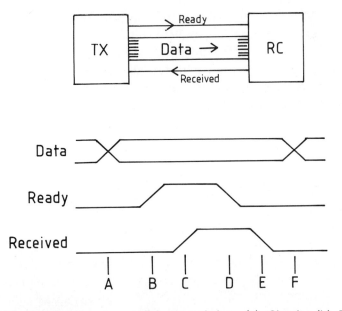

Fig. 11.14 Handshaking on parallel transmission. (a) Circuit. (b) Timing waveforms.

11.5.2 Parallel transmission

The essentials of a parallel transmission channel are shown on fig. 11.14. The timing is achieved by 'handshaking'. At time A the transmitter places new data on the lines. After a small delay to allow the lines to settle the ready line is taken to a 1 at time B. This signals to the receiver that new (and stable) data is present on the lines. The receiver accepts the data (usually by gating it into an eight-bit latch) and signals acceptance by taking the received line to a 1 at time C.

When the transmitter sees the received signal it clears the ready signal at time D. The receiver sees the ready signal clear and clears its received signal at time E. When the transmitter sees the received signal cleared, it can send the next data word at time F.

Handshaking ensures transmitter and receiver stay in step, and allows the receiver to slow down the transmitter if it is busy for some reason. Although it is possible to construct the transmitter and receiver logic with discrete logic, ICs known variously as PIO (parallel input/output), PIA (peripheral interface adaptor) and PPI (programmable peripheral interface adaptor) are available. Devices such as the Intel M8212 are also designed for parallel data transmission, and contain an eight-bit latch, tristate output and control logic.

11.5.3 Serial transmission

One possible form of serial data transmission is shown in fig. 11.15. Data to be transmitted is parallel-loaded into a shift register along with a start bit and a stop bit. Data is then shifted out onto the line one bit at a time. When all eight bits and the stop bit have been sent a ready signal is generated to signal that a new word can be loaded into the shift register.

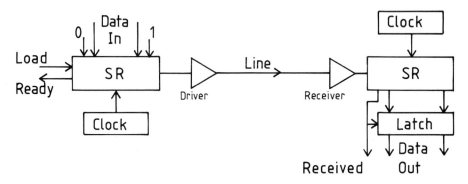

Fig. 11.15 Serial link block diagram.

Data arriving at the receiver is shifted from the line into another shift register. When all bits have been received a 'data received' signal latches data from the shift register to an eight-bit parallel latch and notifies the rest of the receiving device that a new data word has been received.

The data transmission and reception rates are determined by separate clocks. These must match in frequency, and crystal oscillators are usually employed. Alternatively, one clock could be used at the transmitter, say, and two signal pairs used; one for data and one to take the clock to the receiver.

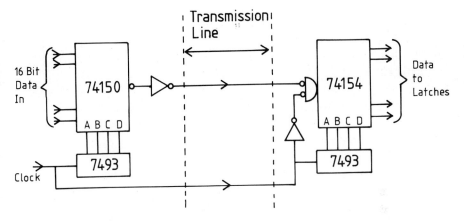

Fig. 11.16 Serial transmission using multiplexers/demultiplexers.

An alternative scheme, shown in fig. 11.16, uses a multiplexer at the transmitter and a demultiplexer at the receiver to achieve parallel/ serial and serial/parallel conversion. Parallel data is presented to a data selector such as the 74150. The address lines are connected to a continuously running counter, so the data is selected one bit at a time and passed serially to the line.

At the receiver, eight RS flip-flops are initially cleared. A 74154 demultiplexer has its address lines connected to another counter operating in synchronisation with the transmitter counter. Received data appears on the correctly selected output and is stored by the corresponding latch.

Serial transmission is often achieved with ICs known as UARTs (Universal Asynchronous Transmitter/Receiver). These incorporate all the control logic and parallel/serial, serial/parallel conversion circuits plus synchronising and error detecting logic. The commonest UART is the 6402.

11.5.4 Error checking and parity
Data transmission is particularly vulnerable to noise from external interference sources, especially when data is carried over telephone lines or radio links. It is usual to use some form of error detection to reduce the effect of this noise.

The simplest form of error checking is the parity bit. As explained earlier, data is usually transmitted in eight-bit words. An extra parity bit is added to the word. The parity bit is set or reset to ensure that the total number of bits in the word plus parity is odd. For example:

Data	Parity
01101101	0
11001001	1
00000110	1
11011111	0

In each case the total number of bits is *odd*; this is known as odd parity. It would, of course, be equally valid to design a system with a parity bit which made the total number of bits *even*; this is known as even parity.

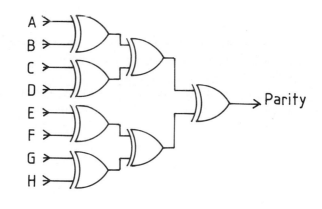

Fig. 11.17 Eight-bit parity tree.

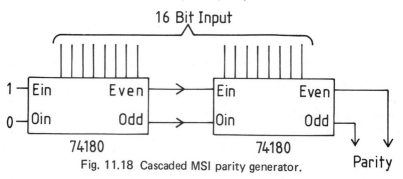

Fig. 11.18 Cascaded MSI parity generator.

Parity in parallel systems can be generated by a tree of XOR gates similar to fig. 11.17. Parity generator ICs such as the 74180 are more commonly used. These can generate odd and even parity, and are provided with odd and even parity inputs to allow parity generation for any desired word length. Figure 11.18 shows two 74180 ICs cascaded to generate parity for a 16-bit word. The equivalent CMOS device is the 4531 13-bit parity generator.

Parity on serial transmission can be generated either on the initial parallel input or by a simple toggle flip-flop which operates on the serial bits going onto the line.

Parity checking can be done with identical circuits to parity generation. The parity of the received data is checked against the received parity. A difference signifies an error. Parity can detect, but not correct, one error only, and is useful where the probability of error is small. Where greater confidence is required, or more noise is likely, more secure methods are used.

The first common technique is the sum check. Data is sent in blocks of, say, 128 words followed by a sum check word which gives the total number of bits sent. This gives error detection, but not correction.

Error correction is provided by a code with built-in redundancy, i.e. more bits are sent than are strictly necessary to carry data. Speech, for example, has considerable redundancy which allows the spoken word to be clearly understood amongst background noise. One commonly used code is based on the chain code generator of section 7.6.7. This allows detection of the original message word with nearly 50 per cent corruption of the transmitted chain code word.

11.5.5 Signals and standards

There are three common standards for serial data transmission (although two of these are sufficiently close to be considered identical). All standards, however, use the format of fig. 11.19. Between characters the signal line is at a 1. A 0 start bit indicates the start of a word. There then follows the data bits (seven or eight according to the application) in true form. The data is followed by a parity bit and one (or two) 1 level stop bits indicating the end of the character. It thus takes ten or eleven bits to send an eight-bit word.

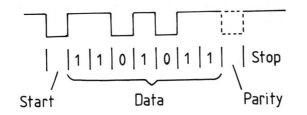

Fig. 11.19 Serial data format.

Data transmission speed is given in Baud which is the number of bits transmitted per second. A 110 Baud link will therefore carry ten characters per second. Speeds of 110, 330, 1200, 2400, 4800 and 9600 Baud are commonly used.

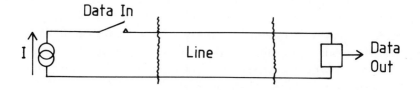

Fig. 11.20 20 mA serial link.

The commonest transmission standard is the 20 mA loop which, oddly, has no official standard. The data is sent as current pulses, with 20 mA representing a 1 and 0 mA representing a 0. A simple link can be represented by fig. 11.20 which has a current source, a data switch (e.g. a transistor) and a current sensor (usually an opto-isolator, see section 13.5.2). Because currents are used rather than voltages, a 20 mA loop has high noise immunity.

One problem caused by the lack of a formal standard is the different convention for the placement of the current source which can, with equal validity, be either at the transmitter or receiver. Figure 11.20 is said to have an active transmitter/passive receiver. A loop with the current source at the receiver is said to have a passive transmitter/active receiver.

The other two standards are very similar, and are known as RS232 (American) and V24 (European). These use voltages to represent the data, with a negative voltage representing a 1 and a positive voltage representing a 0. Usually levels of ± 12 volts are used.

Communication solely from a transmitter to a receiver is called simplex transmission. Bidirectional (but not simultaneous) communication between two points is known as half duplex. Full simultaneous bidirectional communication (usually on two signal pairs) is known as full duplex.

11.6 Communications ICs

11.6.1 TTL

Single-ended line drivers		*Availability*
74128	Quad two-input NOR	N
74140	Dual four-input NAND	S

Both the above devices have sufficient source and sink capability to drive terminated 50 ohm and 75 ohm lines.

Buffers/inverters (high current, open collector, tristate)

		Volts	*mA*	*Type*
7406	Hex inverting buffer o/c	30	40	N
7407	Hex buffer o/c	30	40	N
7416	Hex inverting buffer o/c	15	40	N
7417	Hex buffer o/c	15	40	N
7428	Quad two-input NOR	TTL	48	N
		TTL	24	LS
7433	Quad two-input NOR o/c	5	48	N
	o/c	5	24	LS
7437	Quad two-input NAND	TTL	48	N
		TTL	24	LS
		TTL	60	S
7438	Quad two-input NAND o/c	5	48	N
	o/c	5	24	LS
	o/c	5	60	S
	o/c	15	48	N suffix A
7440	Dual four-input NAND	TTL	48	N
		TTL	60	H
		TTL	24	LS
		TTL	60	S
74125	Quad buffer	Tristate	–	N, LS
74126	Quad buffer	Tristate	–	N, LS
74128	Quad two-input NOR	TTL	48	N
74240	Octal inverting buffer	Tristate	–	LS, S
74241	Octal buffer	Tristate	–	LS, S
74244	Octal buffer	Tristate	–	LS
74365	Hex buffer	Tristate	–	N, LS
74366	Hex inverting buffer	Tristate	–	N, LS

Buffers/inverters (high current, open collector, tristate)

		Volts	mA	Type
74367	Hex buffer	Tristate	–	N, LS
74368	Hex inverting buffer	Tristate	–	N, LS
74425	Quad buffer	Tristate	–	N
74426	Quad buffer	Tristate	–	N

Data selectors/multiplexers and decoders — Availability

		Availability
7442	BCD to decimal decoder	N, L, LS
74150	16-input data selector	N
74151	Eight-input data selector	N, LS, S
74153	Dual four-input data selector	N, L, LS, S
74154	Four-line to 16-line decoder	N, LS
74157	Quad two-input data selector	N, L, LS, S
74158	Quad two-input data selector inverting	LS, S
74251	Eight-input data selector tristate	N, LS, S
74253	Dual four-input data selector tristate	LS, S
74257	Quad two-input data selector tristate	LS, S
74258	Quad two-input data selector inverting tristate	LS, S
74351	Dual eight-input data selector	N
74352	Dual four-input data selector inverting	LS
74353	Dual four-input data selector inverting tristate	LS

Other decoders may be found in sections 8.14 and 9.8.

Data selectors/multiplexers with latch storage — Availability

		Availability
7498	Quad two-input data selector	L
74298	Quad two-input data selector	N, LS
74354	Eight-input data selector	LS
74355	Eight-input data selector, open collector only	LS
74356	Eight-input data selector with D-type	LS
74357	Eight-input data selector with D-type, open collector only	LS
74398	Quad two-input data selector	LS
74399	Quad two-input data selector	LS

Demultiplexers

74137	3 to 8 line with address latch
74138	3 to 8 line
74139	Dual 2 to 4 line
74154	4 to 16 line
74155	Dual 12 to 4 line
74156	Dual 2 to 4 line open collector
74159	4 to 16 line open collector

Transceivers

All **LS** except 74226 which is S

74226	Quad with latch tristate
74242	Quad inverting tristate
74243	Quad tristate
74245	Quad tristate
74440	Quad open collector
74441	Quad inverting open collector
74442	Quad tridirectional tristate
74443	Quad inverting tridirectional tristate
74623	Octal tristate
74640	Octal inverting tristate

11.6.2 CMOS

No CMOS devices can drive terminated lines.

Tristate buffers and transceivers

40097	Hex buffer tristate
40098	Hex inverting buffer tristate
40244	Octal buffer tristate
40245	Octal transceiver tristate

Some latches and flip-flops have tristate outputs (4043, 4044, 40254, 20373, 40374)

Data selectors/multiplexers and decoders

4019	Quad two-input multiplexer
4028	BCD to decimal decoder

4512	Eight-input multiplexer tristate output
4514	1-of-16 decoder with input latches. True output (Can also be used as demultiplexer)
4515	1-of-16 decoder with input latch. Complement output (Can also be used as demultiplexer)
4519	Quad two-input multiplexer
4539	Dual four-input multiplexer

Analog multiplexers are covered in section 12.6.

11.6.3 75 Series devices
These are part of the Texas Instruments 75 Series Analog Family.

Drivers

75150	Dual RS232
75158	Dual RS 422 differential
75159	Dual RS 422 differential
75188	Quad RS 232
75186	Quad RS 432
75172	Quad RS 422 tristate differential
75174	Quad RS 422 tristate differential

Receivers

75152	Dual RS 232 single ended
75154	Quad RS 232 single ended
75189	Quad RS 232 single ended
75173	Quad RS 422/3 differential
75175	Quad RS 422/3 differential

11.6.4 Other communication ICs

Many tristate buffers/latches/transceivers are available as microcomputer-related ICs (e.g. Motorola 6880-6889). ECL devices such as MC10124, MC10125 will act as line drivers and receivers.

Line drivers

1448	RS 232 line driver
9636	Dual RS 423 line driver
9638	Dual RS 422 line driver

Line drivers
MC3487   Quad RS 422 line driver
  3691   RS 422/3 line driver

Line receivers
    1489   RS 232 line receiver
88 LS 120   RS 422/3 differential line receiver
MC 3486   Quad RS 422/3 differential line receiver

CHAPTER 12

Analog Interfacing

12.1 Introduction

There is often a rather 'murky' area when a system contains both analog and digital circuits which interact. The interfacing between analog and digital sections is a topic that is often overlooked. This chapter describes techniques and devices for analog/digital interfacing. In addition, a few miscellaneous devices are discussed which do not fall neatly within the compass of other chapters in this book.

12.2 The transmission gate

Using CMOS techniques it is possible to construct a digitally controlled analog switch. These devices are known as transmission gates, and can be considered as solid state relays. The symbol for a transmission gate is shown in fig. 12.1a. When the control input is 1 the contact is closed, SW1/SW2 are connected and Vo follows Vin. With the control input at 0, the switch is open and Vo is disconnected from Vin.

The on resistance (i.e. the resistance between SW1/SW2 when the contact is closed) is typically 50-500 ohms, dependent on the device. There is *no* offset voltage between input and output. The off resistance is of the order of 100 Megohm. Vin (and hence Vout) can vary between 0 V and the positive supply, V_{DD}.

The commonest transmission gate IC is the 4016 (also called an analog switch). This has four independent gates as shown in fig. 12.1b. The on resistance is typically 300 ohms. An improved version of the 4016 is the 4066. This has the same arrangement as fig. 12.1b, but has a lower on resistance, typically 70 ohms, plus some internal improvements.

If it is required to switch a voltage which can be positive or negative, the 4016 and 4066 can be used with V_{DD} positive and Vss negative, and simple one-transistor level changers on the control inputs.

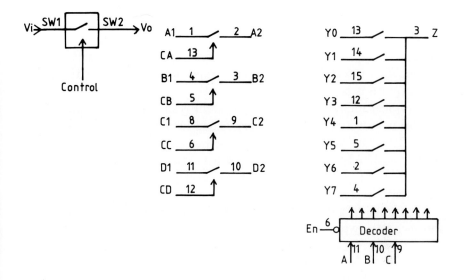

Fig. 12.1 Analog switches and transmission gates. (a) Transmission gate. (b) 4016 Analog switch. (c) 4051 eight-way analog multiplexer.

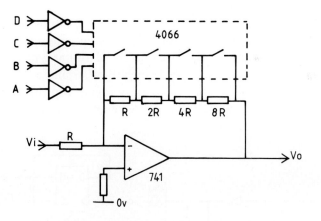

Fig. 12.2 Amplifier with digitally selected gain.

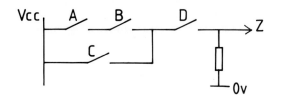

Fig. 12.3 Logic functions with analog gates.

There is a family of transmission gates in the CMOS family, and section 12.6 lists the common devices. These include analog selectors such as the 4051 shown in fig. 12.1c. This is a 1-of-8 selector, with the analog input being selected by three digital inputs ABC. These are decoded internally to close one transmission gate (e.g. if CBA is 110, input 6 is connected to the output).

A typical mixed digital/analog system is shown in fig. 12.2, where a 4066 is used to short out binary scaled feedback resistors on an Op Amp. The binary input ABCD can select integer gains in the range 1 to 15 (with ABCD all 0, the on resistance of four gates will give a gain of about 0.01). The circuit performs as a digitally controlled amplifier.

Transmission gates can also be used to perform logic. Figure 12.3 shows a 4016 wired to produce the function:

$$Z = ((A.B) + C).D$$

Transmission gates can also be used to switch/select analog signals according to digital signal conditions.

12.3 The Schmitt trigger

The Schmitt trigger, described in section 3.2.11, can be used to interface slowly changing signals to fast digital logic. Figure 12.4 shows a typical example, where a digital clock signal running at twice mains frequency is provided by a transformer/rectifier and a 7414 Schmitt trigger. A D-type 7474 connected as a toggle flip-flop divides the Schmitt output frequency by two to give an equal mark/space pulse train at mains frequency.

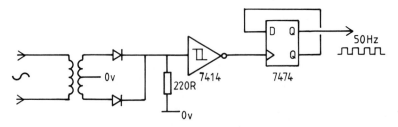

Fig. 12.4 Schmitt trigger with 'analog' input.

12.4 Digital to analog conversion

A binary number can represent an analog voltage. For example, an eight-bit number can represent any decimal number from 0 to 255 (or −128 to +127 if twos complement representation is used). An eight-bit number could, therefore, be used to represent a voltage in the range 0 to 2.55 volts (say) with each bit representing 10 mV. (If twos complement representation is used, an eight-bit number can represent −1.28 volts to +1.27 volts.) Other scalings could of course be used.

A device which converts a digital number to an analog voltage or current is called a digital to analog converter or DAC. Most designers use IC or hybrid DACs, but it is useful to examine the principles of DACs.

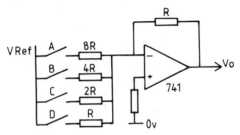

Fig. 12.5 Simple digital to analog converter (DAC).

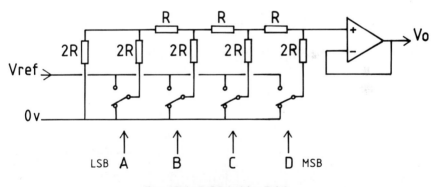

Fig. 12.6 R-2R ladder DAC.

A simple four-bit DAC is shown in fig. 12.5. This uses binary scaled resistors (i.e. resistor values double from R to 8R). These resistors are connected to a stable voltage Vref by transmission gates A to D. Because the circuit acts as an operational amplifier adder, the output voltage is a representation of the binary number ABCD (D is most significant, A least significant) controlling the four transmission gates. Obviously the circuit can be expanded to any required number of bits. Eight bits gives a resolution of 1 part in 255, 10 bits 1 part in 1024 (i.e. 0.1 per cent), 12 bits 1 part in 4096 (i.e. 0.025 per cent).

Another common DAC is the R-2R ladder of fig. 12.6. Transmission gates switch between 0 V and a stable voltage Vref. Analysis of the circuit again shows that the binary number on ABCD controlling the transmission gates is converted to an analog output voltage. The circuit can again be extended to any number of bits. The R-2R DAC is more common because it requires just two values of resistor and is easy to make in IC form. The summing DAC of fig. 12.5 requires an assortment of resistor values and for more than eight bits the least significant resistors need to be in excess of 1 megohm, which is difficult to achieve in IC form.

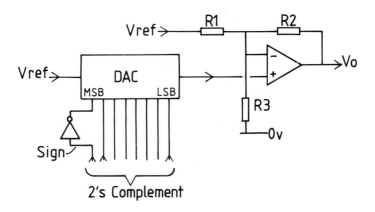

Fig. 12.7 Twos complement DAC.

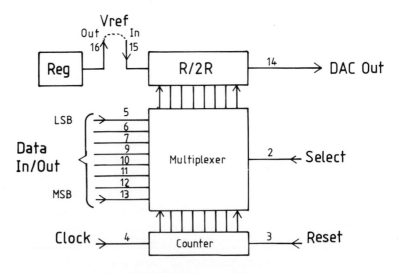

Fig. 12.8 ZN425 counter/DAC.

Twos complement conversion can be achieved by simply offsetting the output of a DAC. Figure 12.7 shows an eight-bit twos complement DAC. R1 is connected to a positive voltage such that the output voltage from the adder is 0 volts with a binary input of 1000 0000 to the DAC itself. The most significant bit to the DAC must be inverted from the logic so a twos complement input of 0000 0000 gives 0 volts out. The circuit will then give an analog voltage which represents the twos complement input, with negative voltages for 1000 0000 to 1111 1111 and positive voltages from 0000 0000 to 0111 1111.

There are no specific DACs in the TTL or CMOS families, but many DAC ICs are available from various manufacturers. Typical is the Ferranti ZN425 shown on fig. 12.8. This is an eight-bit device, and gives an analog output from 0 to 2.55 volts (i.e. 1 bit is 10 mV). The stable reference voltage is obtained from an on-chip regulator. The device also includes an eight-bit counter. The DAC portion can be driven externally (pin 2 select low) or from the counter (pin 2 high). If the counter is selected the data pins D0-D7 give the counter state. The counter is driven from the clock input (pin 4) and can be reset to zero by pin 3. The counter is provided to allow the device to be used as part of a ramp ADC, described further in section 12.5.3.

The accuracy of a DAC is usually specified by its linearity, which defines the maximum deviation of the DAC output from its theoretical value. In most well-designed DACs, the linearity will be 0.5 of the least significant bit. In considering the accuracy of a DAC it should be remembered that any DAC has inherently a finite resolution; an eight-bit DAC has a resolution of 1 part in 256.

Another important factor is the settling time. This is the time taken for the DAC to respond to a change in the digital input, and is specified as the time from the change until the analog output is within 1 bit of the final value. DACs are very fast (the ZN425 has a settling time of 2 μs) but the overall response may well be slower if scaling amplifiers are used after the DAC.

12.5 Analog to digital conversion

12.5.1 Introduction
An analog to digital converter, or ADC, takes an analog input voltage or current and produces a digital representation. It is important to realise that the input voltage can be continuously variable, but the digital output has an inherent resolution. An eight-bit ADC, for example, will only give a resolution of 1 part in 256 or about 0.5

per cent. This resolution (and hence the number of bits) must be chosen with the desired accuracy in mind. There is little point in using an eight-bit ADC in a system requiring 0.1 per cent accuracy.

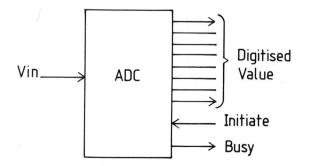

Fig. 12.9 Analog to digital converter (ADC).

A typical ADC can be represented by fig. 12.9. On the left is the voltage to be measured. A conversion usually takes a finite time, so 'handshake' signals are needed between the ADC and the rest of the logic. An initiate signal is used to start the conversion, and a Busy signal shows when the ADC is performing the conversion. When the busy signal clears, the conversion is complete and the digital number may be read on the data line. Inherently, therefore, an ADC takes a 'snapshot' of the input voltage.

This snapshot, or sample, must be repeated at regular intervals. The sample rate must be more than twice the highest frequency in the input signal if a faithful representation is required (as determined by Shannon's Sampling Theorem). A knowledge of the input signal frequency content is needed to choose the sample rate and hence the required ADC conversion time.

An ADC is usually specified by its accuracy (and hence its resolution) and by its conversion time. ADCs are available with conversion times as low as 100 ns and accuracies of better than 0.01 per cent. Three typical ADCs are described in the following section.

12.5.2 The flash converter
The flash converter is the fastest ADC available, but is not used for high accuracy applications because the circuit complexity increases dramatically with the number of bits. Figure 12.10 shows a three-bit converter, with a resolution of 1 part in 8.

The input voltage to be measured is compared simultaneously with seven equally spaced voltages. On fig. 12.10 the input voltage range is

0 to 8 volts, and the comparison voltages are 1, 2, 3, etc. up to 7 volts. If the input voltage is 3.6 volts, for example, the outputs of comparators abc will be high, and all the other comparators will be low.

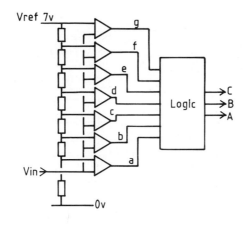

Fig. 12.10 The flash converter.

The seven comparators' outputs are converted to a three-bit binary output by a logic encoder. This is a simple combinational logic network, one possible solution of which is the priority encoders described in section 8.2.1.

The flash converter is very fast, the only constraint being the propagation delays in the various comparators and gates. It cannot be used without some form of control however. If the input is varying, it is possible for invalid transitory states to appear (e.g. 3 to 4 going 011 to 111 to 100, see section 6.9.4). A flash converter is often preceded by a sample and hold circuit (see section 12.5.6).

The main disadvantage of the flash converter is the way the circuit complexity increases dramatically with the number of bits. A four-bit converter requires 16 comparators and a logic encoder about four times as complex as fig. 12.10. Commercial eight-bit flash encoders are available (e.g. the MC10315 designed for TV applications) with conversion times of a few nanoseconds, but are exceedingly expensive.

12.5.3 The ramp converter
A ramp converter is easy to construct, and gives a good compromise between the conflicting requirements of simplicity, accuracy and speed. Figure 12.11 shows the block diagram of a typical ramp

converter.

An eight-bit counter is connected to a DAC, whose output is an analog representation of the counter state. As the counter counts up from zero, the DAC output will be a ramp. The DAC output is compared with a suitably scaled version of the input voltage Vin.

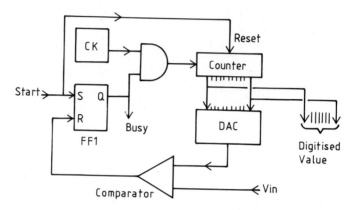

Fig. 12.11 Ramp ADC block diagram.

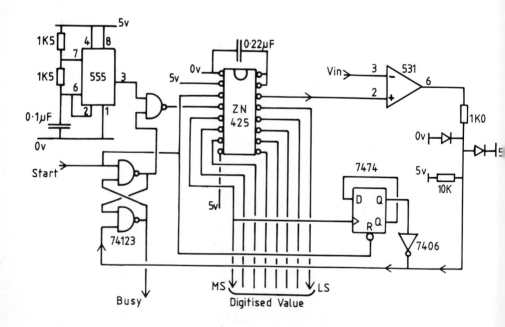

Fig. 12.12 Practical ADC circuit.

To start the sequence a start pulse is given; this sets FF1 and resets the counter. The Q output of FF1 allows pulses from the free-running clock to pass to the counter. As the counter counts up from zero the DAC output ramps up. When the DAC output is equal to the scaled input voltage, the comparator resets FF1. This stops the counter, whose contents are now a digital representation of Vin. The output of FF1 also serves as a digitalisation complete/busy signal.

A practical circuit using the ZN425 DAC is shown in fig. 12.12. The ZN425 incorporates a counter which is linked to the DAC when pin 2 is high. The start flip-flop is provided by the cross-coupled NANDs in the 74132 Schmitt trigger. The 531 Op Amp is the comparator. With the values shown, Vin is measured over the range 0 to 2.55 volts.

If Vin is outside the measurement range, the counter will cycle continuously. This can be prevented by adding a toggle flip-flop driven off the top bit. If the count overspills (indicating Vin over-range) the toggle flip-flop will set and reset the control flip-flop FF1. The toggle flip-flop outputs indicate an over-range input.

The ramp converter is slow because the counter may have to count from zero to full scale. With a 12-bit converter over 4000 clock pulses may be needed to complete a conversion. In addition, the conversion time is not constant, being directly proportional to the input voltage.

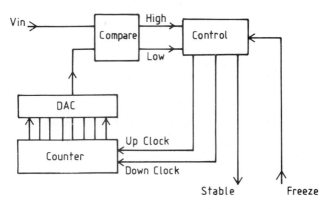

Fig. 12.13 Tracking ADC.

A variation on the ramp converter is the tracking converter. This does not have an initiate signal, but converts continuously. On fig. 12.13 an up/down counter is connected to a DAC. The DAC output is compared to the input voltage by a comparator which has two outputs, an above deadband and a below deadband signal. These control the up/down control lines of the counter. The counter thus tracks the input voltage. A 'within deadband' signal indicates that the counter is stable, and an external freeze signal is provided to prevent the counter changing whilst it is being read.

12.5.4 The successive approximation converter

The successive approximation ADC is a fast, high accuracy device, but is more complex than the ramp ADC described above. The essential features of a successive approximation ADC are shown in fig. 12.14. A DAC output is again compared with the input voltage, but the DAC is controlled by a register rather than a counter. The register outputs are selected by control logic which arrives at the digital representation of Vin by an ordered trial and error process.

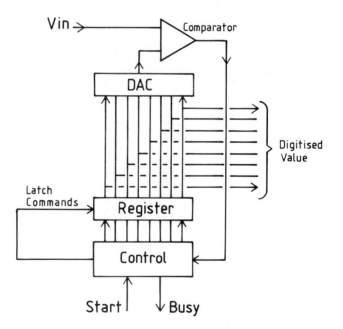

Fig. 12.14 Successive approximation ADC.

The sequence starts with the register cleared. The most significant bit is set, and the comparator output examined. If the comparator shows that the DAC output is less than (or equal to) Vin, the bit is left set. If the DAC output is greater than Vin, the bit is cleared. The next significant bit is now set and tested in the same way. If the DAC voltage is less than or equal to Vin, this bit is left set, otherwise it is cleared again. Each bit is similarly tested in turn until all bits have been tested.

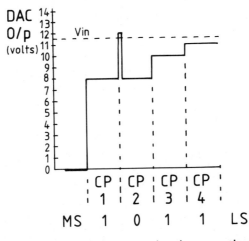

Fig. 12.15 Successive approximation operation.

To illustrate this process, we shall examine a four-bit ADC with an input range of 0 to 16 volts to a resolution of 1 volt, and detail its action in digitising 11.45 volts. The sequence is illustrated in fig. 12.15.

(a) Start of sequence, register cleared.
(b) Clock pulse 1. Bit 3 (MSB) set. DAC output is 8 volts. Comparator shows that DAC output is less than Vin, so bit 3 is left set.
(c) Clock pulse 2. Bit 2 set. DAC output is 12 volts. Comparator shows that DAC output is greater than Vin, so bit 2 is reset again.
(d) Clock pulse 3. Bit 1 set. DAC output is 10 volts. This is less than Vin, so bit 1 is left set.
(e) Clock pulse 4. Bit 0 (LSB) set. DAC output is 11 volts. This is less than the input voltage, so bit 1 is left set.
(f) Digitisation is now complete, and the register holds 1101, the digitised version of the input voltage. Note that the error in the digitisation is 0.25 volts, and is caused by the 1 volt resolution of our 4 bit ADC.

A successive approximation ADC uses the same number of clock pulses as there are bits in the digitised output. A 12-bit ADC (with a resolution better than 0.025 per cent) will convert an analog input to a 12-bit digital representation with 12 clock pulses. A 12-bit ADC has a typical conversion time of 12 μs.

Although it is possible to construct a successive approximation ADC with discrete logic, it is usually more convenient to use one of the specially designed ADC ICs, which contain the registers, DAC and control logic. A typical successive approximation IC is the Ferranti ZN 427, which is shown in block diagram form in fig. 12.16. This is an eight-bit ADC with a conversion time of 15 μs. The device features tristate outputs for use with microprocessor data buses. The input range of the IC itself is 0-2.5 volts, set by an internal zener biased by Rref. If required, an external Op Amp is used for scaling and additional common mode noise immunity.

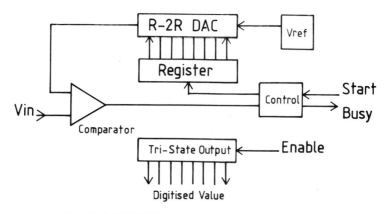

Fig. 12.16 ZN427 successive approximation ADC.

Digitisation is initiated by a 0 start conversion pulse. The end of conversion signal goes to a 0 during conversion, and back to 1 when digitisation is complete.

12.5.5 Bipolar ADCs

All the ADCs described above are unipolar, i.e. they only convert a positive voltage. Bipolar ADCs (which measure positive or negative inputs) are usually constructed by offsetting the input voltage prior to digitisation by a unipolar ADC. For example, an input in the range −5 volts to +5 volts would be offset by 5 volts and measured by a 0 to 10 volts ADC. This is easily achieved by the circuit of fig. 12.17.

To give a true twos complement representation, the MSD from the ADC (which is the sign bit) should be inverted as shown.

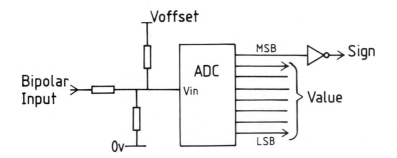

Fig. 12.17 Bipolar ADC operation.

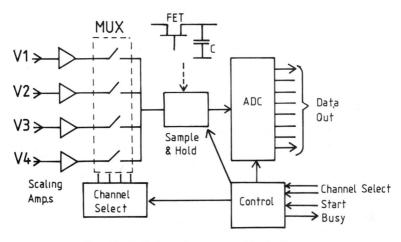

Fig. 12.18 Full analog scanner block diagram.

12.5.6 Analog scanners

When several analog inputs are to be measured, it is often more convenient to use a multiplexed system where inputs are selected in turn and presented to a single ADC. A typical scheme is shown in fig. 12.18, which is a four-input analog scanner.

Each input has its own scaling amplifier. Inputs are selected by analog switches similar to the transmission gates described in section 12.2. The required channel is selected by the decoder logic which decodes the two-bit channel number to four select lines.

The selected channel is 'frozen' by a sample and hold circuit. This freezes the input voltage whilst the ADC performs the conversion and prevents possible ambiguities which can arise with a changing input. The voltage is held on a low leakage capacitor C and the FET is turned off to freeze the signal. The selected voltage is then measured by the ADC, as described in the section above.

It is important when designing analog scanners to know the maximum frequency that an input signal contains. To faithfully reproduce an input, it must be sampled at twice its highest frequency component (Shannon's Sampling Theorem). To illustrate this, a six-input analog scanner will be designed for use with audio signals. These are to be converted to an accuracy of 1 per cent, and have a maximum frequency content of 10 kHz.

An eight-bit ADC has a resolution of 0.5 per cent, so it would seem sensible to use a proprietary eight-bit ADC IC. Each input must be scanned at a minimum of 20 kHz, so the maximum ADC conversion time is $(1/(6 \times 20 \text{ kHz}))$ secs = 80 μs approximately. An eight-bit successive approximation ADC is therefore required to give the required accuracy and frequency response.

12.6 Analog/digital ICs

Transmission gates

4016	Quad transmission gates
4066	Quad transmission gate low impedance

Analog multiplexers/demultiplexers

4051	Eight-channel
4052	Dual four-channel
4053	Triple two-channel
4067	16-channel
4529	Dual four-channel
4551	Quad two-channel

Digital to analog converters

ZN 425	Eight-bit with on-chip counter	1 μs (Ferranti)
ZN 426	Eight-bit R-ZR	1 μs (Ferranti)
ZN 428	Eight-bit with input latches	1 μs (Ferranti)
ZN 429	Eight-bit low-cost R-ZR	1 μs (Ferranti)
MC 6890	Eight-bit current source with input latches	1 μs (Motorola)

Analog to digital converters

ZN 427	Eight-bit successive approximation 15 μs (Ferranti)
ZN 432	Eight-bit successive approximation 15 μs (Ferranti)
ZN 433	Eight-bit tracking, variable conversion time (Ferranti)
MC 10315	Seven-bit flash 50 ns (Motorola)

Analog to digital converters
Many other ICs and hybrid devices are available for analog switching and A/D, D/A conversion.

CHAPTER 13

Practical Considerations

13.1 Introduction

In considering digital systems, it is very easy to get carried away by theory and totally ignore the practicalities of making the device work. Very often a theoretically correct circuit does not work as expected when built, or even worse, exhibits intermittent and unpredictable operation.

This chapter is concerned with often overlooked practicalities such as fault finding, construction techniques, preventing noise, and power supplies. By following the (usually commonsense) procedures outlined below, the designer should be able to design circuits which work in fact as well as in theory.

13.2 Construction techniques

There is no uniformly ideal construction technique, because different applications have different needs. A product being mass-produced will need quick assembly, predictable electrical characteristics and a professional-looking finish. At the other extreme, the designer of a one-off or prototype circuit will look for ease of fault finding and the ability to make design changes.

There are basically four construction techniques available to the designer:
(a) Printed circuit boards (PCBs)
(b) Wire wrap
(c) Wire pen
(d) Stripboards and variants (often called by the trade name Veroboard).

Printed circuit boards need no introduction to most engineers. It is possible to manufacture one-off PCBs with copper-clad boards, etch resist pens and ferric chloride, but this approach is not recommended. Most digital circuits need tracks on both sides of the board, and it is not generally possible to lay out the tracks to the required accuracy with a pen. Assembly is also slow and messy because holes must be drilled individually and, in the absence of plated through holes, solder pins need to be fitted to link sides. Reliability is also poor because slight breaks in tracks (caused by uneven flow from etch resist pens) are quite common. In the author's experience 'home brew' photoresist is even worse.

Professionally produced boards are not particularly cheap. There is a high initial cost to cover the design, artwork and masks, but actual production costs are low. It is generally thought that for an average complex circuit using a double-sided board and plated through holes, the break-even point compared with wire wrap comes between 8 and 12 boards.

Assembly of PCBs is quick, simple and can be done by unskilled (or even automated) labour. Flow soldering, in particular, minimises the commonest intermittent, the dry joint. PCBs, once the design is formalised, should also be free of noise problems caused by, say, cross-talk or supply noise.

Once a PCB is designed, however, changes are expensive. In particular, *retrospective* changes may be impossible. *Ad hoc* changes are unsightly; most manufacturers allow a limit of two to four modifications on a board by track cuts or wire links before a new issue board must be made. It follows that a design must be well proven before a PCB can be considered.

Wire wrapping is a reasonable compromise where a few boards are to be manufactured. Connections are made by wrapping thin single-strand wire round small square pins. Connections can be made to the rear (non-component side) of the board by using IC sockets with extended pins, or on the front (component) side of the board by using separate pins. The typical wire wrap board of fig. 13.1 uses the latter technique. Although slightly more labour intensive (each pin has to be inserted separately rather than with the socket) component side wiring is less error-prone because pin numbering is not reversed.

Wire wrapping works by using the sharp edge of the square section pins to 'cold weld' into the wire, which is itself precisely 'strained' by the wrapping. It is not possible to wire wrap without the correct wire and specialised tools. Correctly done, wire wrapping is very reliable.

There are three forms of wire wrapping. The original, shown in fig. 13.2a, wound solely the wire round the pin. Under vibration it was possible for wires to snap at the first wrap. The modified wrap of fig. 13.2b overcame this problem by wrapping the first two or so turns with the insulation left on the wire. This greatly increases vibration resistance by removing the obvious 'pivot point' where the insulation ends. A recent development uses unstripped wire. The wire, complete with insulation, is wrapped round the pin, whose sharp edges cut through the insulation and into the wire itself.

Fig. 13.1 A wire wrap board.

Wire stripping, wrapping and removal all require special tools. Hand wrapping tools are available relatively cheaply, but their use for large boards is tedious. Power tools, similar to fig. 13.3, are much quicker but rather expensive. Wire wrapping, like most things in life, requires a certain amount of practice. The usual problems are cutting wires too short (which leads to wires being pulled tight round other pins and, in time, to cut through the insulation and cause intermittent faults) or too long (which turns the board into an untidy sprawl). When using the tools, downward force will cause the wire to snap; the wrap tool should just be allowed to rest over the pin.

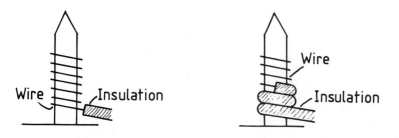

Fig. 13.2 Wire wrapping. (a) Standard wrap. (b) Modified wrap.

Fig. 13.3 A wire wrap tool.

Fig. 13.4 Wire wrapping for easy modification. (a) Difficult to remove. (b) Easy to remove.

Pins will normally take three wires, but it simplifies modifications if connections are made such that each pin has no more than two wires. The obvious way to do this (fig. 13.4a) is not the best. If wire X requires removal for a modification or to correct an error, every wire has to be removed. The arrangement of fig. 13.4b allows any wire to be removed with no more than two unnecessary removals.

The wire pen is a recent and cheap alternative to wire wrapping. Fine-gauge wire coated with heat-sensitive insulating enamel is mounted on a small drum which is itself mounted on a pen similar to a ball point pen, the wire exiting via the tip. The technique needs no special boards, pins or tools beyond the pen and wire. Point-to-point wiring is used, and at each point the wire is given two or three turns round the IC leg or component lead. When soldered, the enamel melts locally giving a good soldered joint. The technique is called Insulation Displacement and is available under the trade names Speedwire or Quick Connect.

Wire pens are quicker than wire wrap because there is no need to cut or strip wires; one chain is done in one go. The resulting circuit is not as easily modified, however, and can look a bit unsightly if not done carefully. If sockets are not used, component replacement necessitates a re-wire. In general, wire wrapping seems to be preferred for commercially produced circuits, but the wire pen does seem to be an excellent cheap alternative for the construction of prototypes.

Stripboards are not really suitable for digital circuits except for amateur and prototype use. Stripboards designed for dual-in-line ICs should be used, and the best layout uses power supply tracks laid out as in fig. 13.5. This arrangement simplifies decoupling and minimises supply noise. Single-strand wire should be used, but a steady hand for soldering is needed if solder bridges are to be avoided.

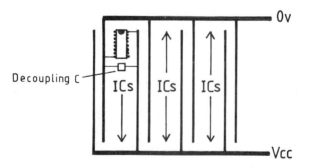

Fig. 13.5 Power supply tracks on stripboards.

Whatever method is used, due note should be made of the precautions outlined in chapter 11. ECL and Schottky, in particular, do not perform well with breadboard construction. High speed ECL should really be used with multi-layer boards (constructed as in fig. 13.6) to minimise power supply inductance and provide stable characteristic impedances.

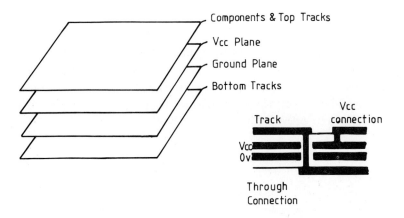

Fig. 13.6 Multi-layer board and plated through holes.

Views differ on the advisability of mounting ICs in sockets. Certainly early sockets were expensive and a constant source of intermittent faults. Recent designs are much improved and are now cheaper than the ICs themselves. Sockets should definitely be used at the prototype stage, and should be considered for production boards unless high vibration is likely to be encountered. The use of low profile sockets (where the IC sits inside the socket) should be avoided as these preclude the use of test clips.

There has recently been a trend away from the traditional board connection, with the gold-plated board edge fitting into an edge connector. These have always been troublesome and prone to tarnishing, even with gold plating. Many intermittent faults can be cured by removing the boards and running a rubber over the board edges.

Greatly improved reliability is obtained by the use of two-part connectors with proper pins and sockets. The board of fig. 13.1, for example, uses a locking Berg connector.

Boards should be mounted to minimise dust collection and promote air circulation for heat removal. TTL and ECL, in particular, generate a surprisingly large amount of heat. Fans are noisy, unreliable and spread dust; natural convection should be used wherever possible.

13.3 Noise

Noise in a digital circuit causes intermittent and unpredictable operation, and it can be very difficult to identify a cause and effect a cure. It is by no means unknown, for example, for prototype boards to behave perfectly and final production boards to be plagued with noise problems. The actual noise immunity of digital circuits was described in section 1.5.4. In this section common causes of noise, and methods of prevention, are described.

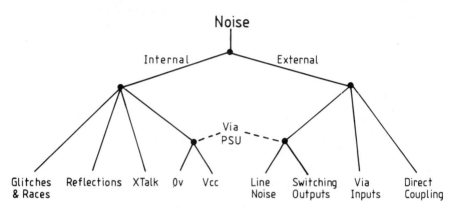

Fig. 13.7 Common causes of noise.

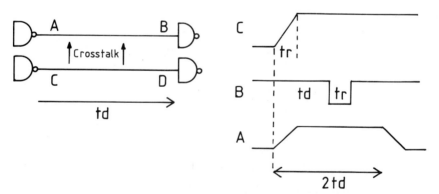

Fig. 13.8 Crosstalk between adjacent lines. (a) Circuit. (b) Waveforms.

Noise can be considered to be the result of all the branches of fig. 13.7, which demonstrates that there is no one cause of noise. Often noise problems arise from combinations of several circumstances.

Internally generated noise is a function of fast transition times. Edge times are totally independent of a circuit's operating frequency, so it is quite possible for a circuit operating at very low speeds to have noise problems. It follows that CMOS, with its slow edge speeds, is relatively free from internally generated noise, but particular care needs to be taken with Schottky TTL and ECL. Glitches and races should not occur in a well-designed system and are not strictly noise. Their symptoms are similar to noise however. Prevention of these problems was described in earlier chapters.

Reflections caused by transmission line effects were covered in chapter 11. Crosstalk is a related effect, and occurs where two signal lines run parallel to each other as in fig. 13.8. Cross-coupling of signals between the two lines will occur, caused by mutual inductance and capacitance of the lines.

It is possible to analyse crosstalk mathematically, but usually the relevant line impedances and mutual inductance/capacitance are not accurately known. Figure 13.8b shows the form of crosstalk predicted by the theory for an edge from gate C.

At point B, a narrow pulse appears equal in duration to the rise time of the signal. This is known as forward crosstalk. The amplitude is directly proportional to the line length and inversely proportional to the rise time (i.e. the faster the edge, the greater the crosstalk amplitude).

At point A, a broad pulse appears equal in duration to twice the line propagation time. This is known as backward crosstalk. The amplitude is determined solely by the line characteristics and is independent of line length and signal rise time.

Crosstalk is exacerbated by unterminated lines. Figure 13.8a is terminated at the receiving end, but not at the transmitting end. A reflected version of the backward crosstalk would therefore appear at point B as well as the forward crosstalk.

Crosstalk can be avoided by minimising long parallel runs or using gates with slow edges. In general, if the line length rules outlined for transmission lines are followed, crosstalk should be minimal. If long parallel runs are unavoidable, crosstalk can be minimised by increasing line separation or by using twisted-pair or coaxial cable with line drivers and receivers.

Power supply noise can be introduced into a system from external sources or generated internally by the system itself. As gates switch from one state to another, their current drain from the supply will change. Fast edges mean current changes. Gate outputs will also charge and discharge stray capacitance which will again cause current pulses to be drawn from the supply. In addition, TTL variants have the slightly undesirable characteristic of drawing a large current pulse during switching. All these effects add together to produce power supply noise generated within the system.

Current changes produce power supply voltage variations by supply line inductance and resistance. In most systems lead inductance is the dominant factor. For a current change Δi in time Δt with lead inductance L, the resulting noise voltage will be:

$$V = L \frac{\Delta i}{\Delta t}$$

Typical values for a TTL gate would be a 20 mA change in 5 ns with a lead inductance of 0.1 μH (corresponding to about 15 cm of PCB track). These values predict a noise pulse of 0.4 volts. It should be remembered that current flows out of the supply rail and into the 0 volt rail. A current change will produce reinforcing noise pulses on each rail. The above current change would remove all of a TTL gate's noise immunity. With several gates switching, the system would probably be swamped with its own noise.

Power supply noise cannot be cured by the obvious solution of a strategically placed electrolytic capacitor as the problem is caused by local lead inductance. An even distribution of small decoupling capacitors is required. Disc ceramic capacitors of 0.01 μF value are used at a density of one capacitor per one to five ICs, dependent on family and sensitivity. A one to one ratio is usual in commercial circuits. The capacitors are, fortunately, cheap.

For a capacitor C and current change Δi in time Δt the resulting voltage change is:

$$V = \frac{\Delta i \times \Delta t}{C}$$

Substituting i = 20 mA, t = 5 ns (as before) and C = 0.01 μF gives a noise voltage of 10 mV; negligible for most logic families.

Ground-borne noise can also be generated by external sources, and is usually caused by poor wiring layout. High currents should, under no circumstances, share return paths with logic. Most designers appreciate the need to separate 0 V returns where digital circuits drive relays or solenoids, but overlook currents from indicators and line drivers. A 5 volt step into a 50 ohm line generates a 100 mA current step.

A sensible ground layout is shown in fig. 13.9. A common earth point (CEP) is established at some point in the cubicle. This is simply a copper busbar which is earthed and is near to the cubicle power supplies. From this CEP, typically four grounds will be established. A hardware ground will earth cubicle doors, frames, etc. for safety. A dirty ground will be connected to high currents from relay, solenoids, etc. A clean ground will be used for the digital circuits and the analog ground for ADCs and similar devices. Note that one PCB could have three different 0 Vs on it, which only connect off the board.

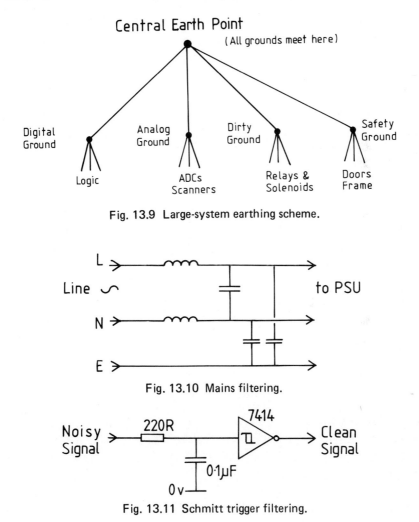

Fig. 13.9 Large-system earthing scheme.

Fig. 13.10 Mains filtering.

Fig. 13.11 Schmitt trigger filtering.

External noise can be carried on the mains voltage feeding power supplies and coupled onto supply rails. Thyristor drives are notorious for causing noise problems in adjacent equipment via the supplies. Conventional decoupling of the power supply output has little effect because of the relatively high inductance of electrolytic capacitors.

The best cure is to identify, and suppress, the noise at source. If this is not possible, a mains filter similar to that shown in fig. 13.10 can be used on the supply input. In extreme cases the whole system can be 'floated' on a battery which is trickle-charged off the mains.

External noise can also enter via inputs and output connections. This can best be prevented by the use of opto-isolation, described further in section 13.5. A simple, but less effective, solution where noisy inputs are present is to use simple RC filtering with a Schmitt trigger as in fig. 13.11. It is, however, good practice never to take the logic 0 V rail outside the cubicle and opto-isolation of inputs and outputs should be used wherever possible.

Extreme external noise can couple directly into the tracks or wiring of a logic circuit by capacitive or mutual inductance effects. Usually this problem arises when the logic board is close to solenoids or relays (in electromechanical printers, for example). The best solution is, again, to suppress the noise at source by means of spike suppression diodes on DC coils, Metrosils on AC coils and RC snubbers across contacts. If thyristors are the cause of the noise, zero voltage switching will reduce noise. If all else fails, the entire circuit may have to be screened by a mu-metal case.

Noise can be exceedingly difficult to locate, and as mentioned earlier, is usually the result of a combination of factors rather than one single cause. Tracing noise requires common sense, patience and a certain amount of trial and error. If noise is expected to be a problem, good noise immunity should be designed-in with opto-isolation and similar techniques rather than added later at the testing stage.

13.4 Mixing logic families

13.4.1 Introduction
Designers tend to be rather narrow-minded and design circuits with one logic family (most designers tend to be very conservative and design *all* their circuits with the family they know best). There is no reason why logic families cannot be mixed, and there are often advantages in, say, mixing TTL and CMOS in one circuit. A portable VHF frequency meter could use ECL for the front-end high speed counters, TTL arithmetic chips, and CMOS general logic to keep the battery consumption reasonable.

This section outlines the circuits commonly used when logic families are mixed. The circuits are grouped according to the sending family.

13.4.2 *From TTL to CMOS*

There are two possibilities: CMOS on a 5 volt rail and CMOS on some other supply. In both cases there is no fan-out problem because CMOS input impedance is negligible compared with TTL.

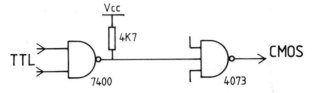

Fig. 13.12 TTL to CMOS with common 5 V supply.

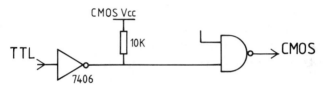

Fig. 13.13 TTL to CMOS with different supplies.

When TTL and CMOS share a common 5 volt supply, it is just possible to drive a CMOS input directly from a TTL output. A CMOS gate recognises a signal above 2.6 volts as a 1 and below 2.4 volts as a 0. TTL 1 output levels are typically 3.4 volts but can fall as low as 2.4 volts when loaded. If a 4K7 resistor is connected from the gate output as in fig. 13.12 the TTL 1 level will rise to almost 5 volts allowing TTL signal to drive a CMOS gate, with no loss of noise immunity.

If the CMOS gates run on some higher supply, the TTL output levels cannot be used directly. The simplest solution is the use of an open collector TTL inverter, such as the 7406 shown in fig. 13.13, with a pull-up resistor to the CMOS supply. The pull-up resistor can be any reasonable value, and should be chosen as a compromise between speed (stray capacitance) and power consumption.

13.4.3 *From TTL to ECL*

There are again two possibilities, ECL and TTL sharing a common 5 volt supply and the more usual case where TTL is on a positive 5 volt supply and ECL is on a negative 5 volt supply.

TTL and ECL have slightly different logic levels and, where they share a common supply, interfacing can be achieved by the simple three-resistor interface of fig. 13.14a. This attenuates the TTL output and shifts it slightly positive to simulate ECL levels.

324

In the more common case of a split supply, it is best to use standard TTL/ECL translator ICs. Typical of these is the Motorola MC10124 quad translator shown in fig. 13.14b. This gives true and complement outputs at ECL levels. (The MC10124 and its equivalent MC10125 ECL to TTL translator can actually be used as TTL line driver/receivers.)

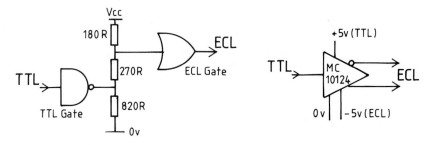

Fig. 13.14 TTL to ECL. (a) Common positive supplies. (b) Separate supplies.

13.4.4 From CMOS to TTL

The basic problem with driving TTL from CMOS is the high sink current required when the signal is at a 0 state. If CMOS and TTL share the same supply, one standard CMOS output can drive one LS or L gate input. Standard CMOS cannot drive standard or Schottky TTL.

The CMOS family contains the 4009 (inverting) and 4010 (non-inverting) buffers. They have a sink capability of 6 mA in the 0 state (whilst keeping the 0 level below 0.8 volts) and as such they can drive about three standard TTL inputs. These devices have separate supply connections for input and output stages allowing them to be used directly between CMOS and TTL on different supplies as in fig. 13.15.

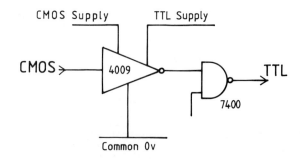

Fig. 13.15 CMOS to TTL using 4009 buffer.

13.4.5 From CMOS to ECL

Translation from CMOS to ECL is not commonly required, except in applications where ECL uses MOS memories. The TTL translator circuits above will work with CMOS although the MC10129 has a higher input impedance and will not load the CMOS output.

13.4.6 From ECL to TTL

In the usual circumstances of a split supply, the simplest solution is to use a ECL/TTL translator such as the MC10125 shown in fig. 13.16a. The MC10125 can drive one standard TTL load.

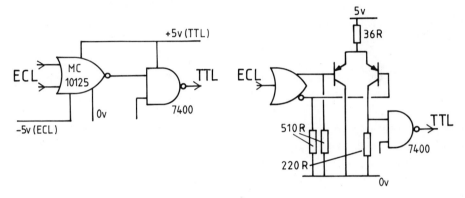

Fig. 13.16 ECL to TTL. (a) Separate supply using MC10125. (b) Common supply with discrete components.

There is no simple solution to an ECL/TTL interface on a common supply. There are no ready-made ICs, and the small ECL swing needs amplification to TTL levels. One possible circuit using a differential amplifier is shown in fig. 13.16b. Another common solution is an analog comparator with one input biased to the midpoint of the ECL levels.

13.4.7 From ECL to CMOS

The circuits in the previous section will interface equally well to CMOS. The resistor values in fig. 13.16b can be much higher in value for CMOS because of the higher impedance levels.

13.4.8 General observations

All the common interfaces have been covered above. When designing less common interfaces, the following criteria should be considered:
(a) Logic level outputs of source, logic level inputs of load and the tolerances on these levels.

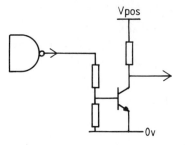

Fig. 13.17 General-purpose buffer (values to suit input/output levels).

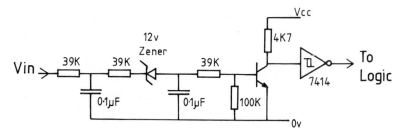

Fig. 13.18 Interface for push-buttons, switches, etc.

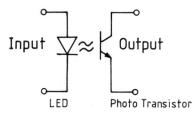

Fig. 13.19 Opto-isolator construction (TIL107).

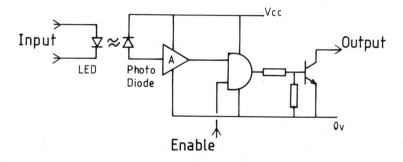

Fig. 13.20 Opto-isolator with amplifier (6N137).

(b) The source current driving capabilities in both logic states.
(c) The load current requirements in both states.

With these considerations in mind, simple inverting buffers similar to fig. 13.17 can be built with discrete components.

13.5 Interfacing to the outside world

13.5.1 Introduction

To be useful, any logic system must communicate with the outside world. Inputs must be obtained from switches, keyboards, contacts and such, and outputs provided to drive displays, relays, solenoids, etc.

Section 13.3 showed that noise is often introduced via inputs and outputs. There is also the possibility that some external fault could introduce mains voltages (or higher) into the system via input or output lines. A well-designed digital circuit will tolerate noise and external faults. This section considers the design of input/output circuits.

13.5.2 Inputs

A very simple input filter was shown previously in fig. 13.11. A more sophisticated circuit which can be adapted to suit any level of input signal is shown in fig. 13.18. With the values shown, it will operate on an input switching between 0 and 24 volts.

Circuits such as those in figs 13.11 and 13.18 work in a reasonable environment, but have the inherent hazard of a direct electrical connection between the circuit and the outside world. Devices called opto-isolators provide almost total isolation and perfect noise immunity.

An opto-isolator consists of an LED and photo transistor in a common encapsulation as shown in fig. 13.19. As current passes through the LED, the emitted light causes current to flow through the transistors. The absence of an electrical connection gives an isolation voltage of 1 kV between input and output, and the ability to 'float' the input gives excellent immunity to common mode noise. The LED operates at a fairly high current (typically 10-20 mA) so there is a good immunity to series mode noise.

A simple opto-isolator (such as the popular TIL 107 or TIL 108) is characterised by a current transfer ratio. This is simply the ratio of output current through the photo transistor to the current through the LED. This is typically 20 to 30 per cent so 20 mA input current will give about 4 mA output current. Transfer ratios as high as 120 per cent can be obtained with Darlington photo transistors, but the voltage drop across the transistor becomes rather large.

More complex opto-isolators (such as the 6N137) incorporate a photodiode, amplifier and output stage as in fig. 13.20. Such devices do not have a transfer ratio as such. Note the useful enable input.

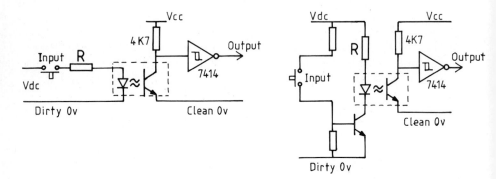

Fig. 13.21 Optical isolation of inputs. (a) Direct switching. (b) Transistor switching.

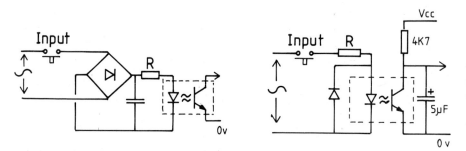

Fig. 13.22 Opto-isolators with AC inputs. (a) Filtering before opto-isolator. (b) Filtering after opto-isolator.

The current through the LED needs to be defined by a resistor. A contact can be used to drive the input directly, as in fig. 13.21a, or at lower contact current, as in fig. 13.21b. In each case the series resistor R defines the LED current as outlined in section 9.2.1. Most opto-isolators require 20 mA.

AC signals can be handled by opto-isolators in a variety of ways. At some point, however, some filtering will be required to turn the rectified AC to a smooth level. This can be achieved before the opto-isolator, as in fig. 13.22a, or after, as in fig. 13.22b. An LED has a very low peak inverse voltage, so the circuit of fig. 13.22b needs a protection diode across the input of the opto-isolator. AC isolator ICs such as the HCPL 3700 are also available.

The circuits above show inputs derived from contacts. Similar ideas can, of course, be used to take inputs from transistors or even other logic gates. Opto-isolators are very fast, and allow data transfer at speeds over 1 MHz if high speed devices such as the 6N137 are used.

Contacts, from relays, switches or whatever, exhibit a phenomenon called contact bounce. A contact does not make instantly, but bounces off and on several times for about 2 to 4 ms. Similar bounce occurs when a contact opens. Even the supposedly bounce-free mercury-wetted contacts are not totally immune. In many applications this does not matter, but it can cause problems in counting or arithmetic circuits.

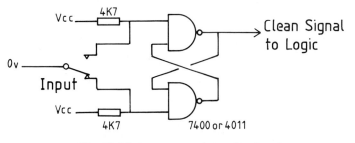

Fig. 13.23 Bounce-removing flip-flop.

Filtering with RC circuits as shown in fig. 13.11 or with retriggerable monostables can provide a simple cure, but the best solution is a changeover contact and a bounce-removing flip-flop as shown in fig. 13.23. Provided a break before make switch is used, this circuit gives totally bounce-free true and complement outputs. The circuit can, of course, be easily modified to use changeover contacts and two opto-isolators. Simpler (but less efficient) circuits are shown in figs 4.9 and 4.10.

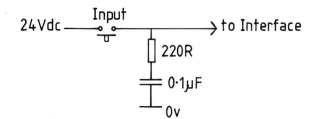

Fig. 13.24 Contact cleaning circuit.

Contact themselves can cause problems. Ideally, contacts should operate at voltages of about 24 volts and currents above 100 mA. These conditions create a small arc as the contacts break, which burns off contamination. At low currents and voltages contacts tend to accumulate a thin insulating film which first gives intermittent operation (and increased contact bounce) and eventual failure. This problem can be overcome by the use of 24 volt supplies and a 0.1 μF capacitor between the switched side and common, as in fig. 13.24. The charge current of the capacitor generates a small arc during the switch bounce, which helps to keep the contacts clean. The best solution, though, is to use sealed contacts or reed relays.

13.5.3 Outputs

Wherever possible, optical isolation and separate supplies should be used on outputs. A typical arrangement using a 7406 to drive an opto-isolator is shown in fig. 13.25. This current can be used (with suitable values of resistor) for any DC load.

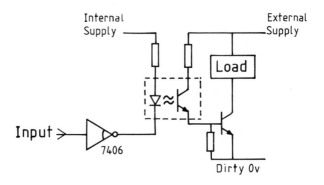

Fig. 13.25 Optically isolated output.

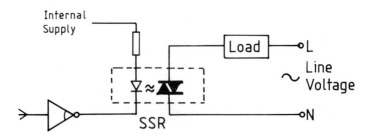

Fig. 13.26 Optically isolated triac (SSR).

AC loads can be switched via an interposing DC relay driven as shown in fig. 13.25, but a more elegant solution is the use of optically isolated triac and thyristor modules. These incorporate optical isolator and triac firing circuits and allow AC loads of up to 40 A to be switched directly from a logic signal. They are ideal for driving AC relays, solenoids, heaters and the like. A possible circuit is shown in fig. 13.26.

These devices are often called solid state relays (or SSRs). The modules are designed to turn on as the AC input *voltage* passes through zero and to turn off when the load *current* passes through zero. This admirable characteristic minimises supply-borne noise. If an SSR is mounted on a logic board, tracks carrying mains voltage should be covered so that there is no personnel shock hazard. Warning labels should also be fitted as service engineers do not expect to find voltages above 15 volts in logic systems.

Inductive DC loads such as solenoids, relays, etc., can generate voltage noise spikes at turn off. The voltage across an inductor is given by:

$$V = L\frac{di}{dt}$$

A fast turn-off generates a high inverse voltage across the coil. This very large voltage spike with a fast rising edge may generate noise, and even damage the output stage. The problem can be simply overcome by the provision of a spike suppression diode as shown in fig. 13.27 (the term free-wheel diode is sometimes used).

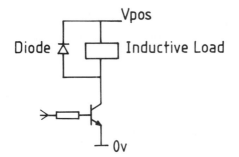

Fig. 13.27 Spike suppression diode with inductive load.

Optical isolation should be used wherever possible, but it is possible to drive small DC loads directly from TTL or CMOS. Circuits capable of driving small relay loads are shown in fig. 13.28. In each case, the relay supply should be separate from the logic supply, and the 0 V return for the load current should only meet the logic 0 V at a central earth point.

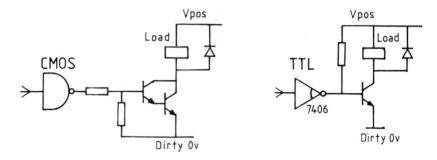

Fig. 13.28 Interfacing directly to small loads. (a) CMOS. (b) TTL.

13.6 Maintenance and fault finding

13.6.1 First principles

Equipment reliability and availability are determined by two factors. The first, and obvious, factor is how often the system fails. This is measured by the Mean Time Between Failure (MTBF). For most systems this will be over ten thousand hours. The second, and not so obvious, factor is how long it takes to repair, measured by Mean Time to Repair (MTTR). Most engineers appreciate the need for a high MTBF, but the necessity of designing in a low MTTR is often overlooked. Both MTBF and MTTR are equally important if high availability is to be obtained. Availability is, in fact, defined as:

$$\text{Availability} = \frac{\text{MTBF}}{\text{MTBF} + \text{MTTR}}$$

MTBF is determined by sensible equipment design: conservatively rating components; use of good quality connectors; avoiding fans; and similar commonsense precautions. It should be remembered that most failures are mechanical in origin, bad connections, breaks in PCB tracks, bad solder joints and the like. It is surprising how many intermittent faults have been designed *into* equipment.

Low MTTR is again obtained by careful design and common sense. Most fault finding is done at two levels. First-level fault finding, often done by technicians, is concerned with board or module changing. The second level repairs faulty boards down to component level.

At each stage, careful design will minimise the repair time. Good, well-thought-out documentation is essential. Many logic diagrams resemble rats' nests or car wiring diagrams. Documentation and drawings should assist the user, not be obscure intelligence tests. The inclusion of status LEDs at strategic points in the circuits adds little to the design effort or cost, but is of great diagnostic use.

If a system size means several PCBs must be used, the designer should carefully consider the allocation of logic to the boards. The easiest systems to repair are those where the repairer can identify functions with boards (e.g. the VDU driver is on board 3). The worst systems are those where one function wanders over several boards. If a sensible layout is adopted, first-level fault finding can be done with block diagrams rather than full circuits.

The sensible user can also reduce the MTTR by ensuring drawings and spares are available and up to date. Spares should be tested at least once; too often the first time spares are tried is when the system fails for the first time. If the spare is faulty chaos ensues. Adequate instruments are essential; it is not possible (despite some firms' apparent belief) to design and repair digital circuits with a multimeter.

13.6.2 Fault-finding aids

Classically, there are two fault-finding techniques: signal tracing and signal injection. The first assumes correct input conditions, then the signal is traced through the circuit until a fault is located. Signal injection uses a monitoring device at the output, and a signal source to inject a signal into the circuit starting at the output and working back until the fault is found. Signal tracing is used almost exclusively in digital circuits.

The commonest fault-finding aid is the oscilloscope. Most engineers are familiar with oscilloscopes, but there are a few points to consider when choosing a scope for digital circuits. Usually the engineer will be interested in time relationships between different parts of the circuit, so a twin-beam scope is a necessity. If it is required to examine glitches and transmission line effects the vertical amplifier rise times must be at least as fast as the rise times of the circuit being examined.

A very useful feature is a dual timebase which allows 'A delayed by B' operation. This feature allows the scope to be triggered by an event, then some operation, say 10 ms later, to be examined on a fast timebase. Figure 13.29 illustrates the use of this feature.

There are many useful variations on a basic oscilloscope. Storage scopes allow a transient event to be frozen. A conventional storage scope only allows the current picture to be stored. A recent development is the digital storage scope (DSO), so called because it uses digital techniques to store the display. A DSO stores the signal in a manner similar to an analog shift register. At any time the contents can be frozen and the entire signal examined at leisure. The display is not limited to one frame, so the signals leading up to an event can be frozen and observed.

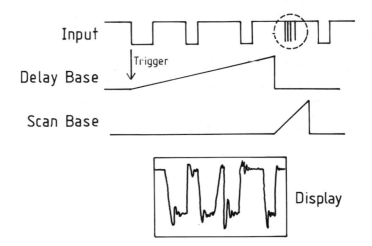

Fig. 13.29 Oscilloscope with delay feature.

Oscilloscopes designed specifically for digital applications often incorporate word triggering. Parallel word triggering will trigger the timebase when some preset bit pattern occurs on the word trigger inputs. Usually eight- or 16-bit word triggering is provided. Serial word triggering is similar, but looks for a preset serial bit pattern on a single trigger input. Word trigger scopes provide selectable parallel/serial triggering. Combined with a DSO they make a very powerful tool for tracing circuit maloperation at the design stage.

The ultimate fault-finding tool is the logic analyser, an example of which is shown in fig. 13.30. In its simplest form this operates as a word triggered oscilloscope, but displays logic levels as idealised traces, not a true analog display. An average logic analyser will, however, display up to 32 inputs at once. The traces are produced by circuits which can simulate TTL, CMOS or ECL characteristics, and

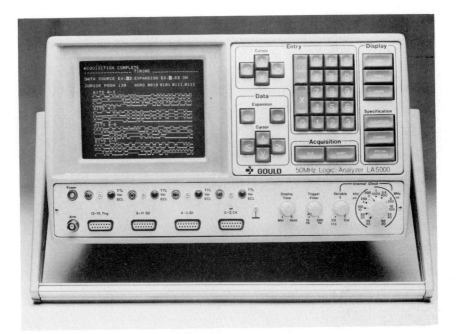

Fig. 13.30 A logic analyser.

Fig. 13.31 A logic probe.

are sufficiently fast to show glitches, races and reflections.

More complex logic analysers have many additional features: ability to display channels in hex, octal or decimal; comparison of a circuit's operation with a pre-stored ideal (signature analysis); and sequence triggering (a more advanced form of word triggering). The price inevitably rises with these extra facilities.

It is not necessary, however, to spend a fortune on fault-finding aids. If a relatively simple fault is being located in a previously working circuit, there are many cheap fault-finding tools.

A logic probe similar to fig. 13.31 allows the state of one signal to be examined on LEDs. Most include a pulse stretcher to allow the presence of a short pulse to be seen. Some cheap logic probes do not, however, distinguish between a logic 1 and a floating (disconnected) input.

Fig. 13.32 A logic checker.

Possibly more useful is the logic checker shown in fig. 13.32, which allows the states of all pins of a 14- or 16-way IC to be examined at once. These ingeniously select the correct pins for Vcc and 0 V themselves. Their one shortcoming is their inability to be used on mixed analog/digital ICs such as timers. A related device is the logic clip of fig. 13.33, which makes attachment of scope and meter leads a straightforward operation. Both these devices require a certain amount of clearance around the IC, and cannot be used on very high density boards, or with low profile sockets.

Signal injection is not common, but logic pulsers are available (similar in appearance to the logic probe of fig. 13.31) which inject short pulses into circuits (overriding gate outputs). These pulses can be traced elsewhere. It is also worth noting that a TTL output can be shorted to 0 V and CMOS to either rail for short periods without harm, which allows signals to be forced for checking.

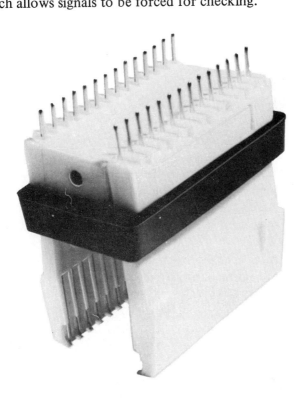

Fig. 13.33 An IC test clip.

If sockets have not been used, the engineer will eventually be faced with the need to remove an IC. This should only be attempted when it is certain that the device has failed; ICs are very reliable and there is always the likelihood of damaging the PCB tracks; plated through holes are particularly vulnerable.

ICs should *never* have all their leads bent over under the board; this makes removal exceedingly tricky. The only way to remove such an antisocially mounted IC is to snip each lead close to the IC body, then desolder each remaining stub lead with a solder sucker. Attempts to straighten the leads with the soldering iron tip will probably damage plated through holes and the PCB track.

The best way to remove ICs requires two people and a special soldering bit which applies heat to all pins of the IC simultaneously. One person applies the bit to the pins on the solder side of the board. The other waits for the solder to melt, then removes the IC with an

IC extractor tool. This lifts the IC cleanly from its board with no damage to the PCB or plated through holes. This method only works with ICs which have had no more than two leads bent over.

13.7 Power supplies

Logic systems require some form of power supply, and in the case of TTL and ECL systems these will be fairly substantial units providing several amps at a tolerance of 5 per cent.

Power supplies for CMOS are relatively easy to provide. CMOS will operate from 5 to 15 volts, and the supply does not even have to be well regulated provided that one supply feeds the whole system. CMOS has very low power requirements, but as explained in section 1.5.2, the current rises with operating frequency. A rough estimate for a system operating at 1 MHz and a 12 volt supply can be obtained by assuming 2 mA for a simple (e.g. gate) package and 5 mA for an MSI (e.g. counter) package. In many systems the input/output interfacing will take more current than the rest of the logic. In general, CMOS current requirements are directly proportional to the operating frequency.

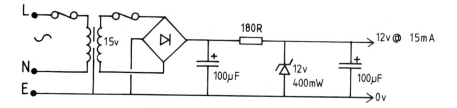

Fig. 13.34 Simple supply for small CMOS system.

Small low-frequency CMOS systems can be used with very simple power supplies. Figure 13.34 will supply about 50 ICs operating at below 50 kHz. The load current will be about 15 mA. Larger systems are best fed from a power supply based on an IC regulator such as the 78 series, which are available in 100 mA, 1 A and 5 A versions and 5, 12, 15 volt ratings. A circuit for a 1 A, 5 volt PSU is shown on fig. 13.35.

CMOS systems often run on batteries, and some form of prediction of battery life will be required. A battery's capacity is given in ampere hours, which is simply the product of load current and the expected life. Rechargeable batteries have a quoted capacity; a PP3 Nicad, for example, has a capacity of 0.1 Ah.

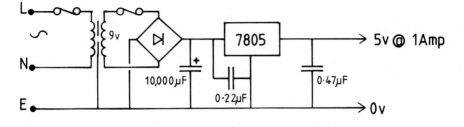

Fig. 13.35 IC regulator supply for TTL or CMOS.

Dry batteries are less predictable, but typical figures are:

Battery	Voltage	Capacity (Ahr)	Range (mA)
PP3	9	0.4	10
Penlight 'AA'	1.5	0.9	30
HP 11 'C'	1.5	1	150
HP2 'D'	1.5	2	250

These figures are based on intermittent use and are only rough guides; there is a wide variation between supposedly identical batteries.

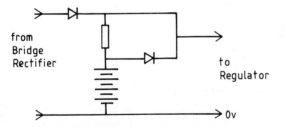

Fig. 13.36 Uninterruptable supply.

Where mains/battery operation is needed it is often convenient to provide rechargeable batteries trickle-charged off the mains-driven power supply. Figure 13.36 shows a possible circuit for battery backup of a CMOS circuit. This features automatic changeover with no interruption on removal of the mains supply.

TTL and ECL are far more demanding, needing high current and tight tolerances. TTL requires 5 volts ± 0.25 volts and will be damaged if the supply rises above 7 volts. The supply must be retained within these limits despite mains variations, cable drops and similar disturbances. Standard TTL current requirements do not vary greatly with speed, and can be estimated for each standard IC:

Gate packages	10 mA
Decoders	20 mA
Flip-flops	25 mA
MSI	60 mA

LS and L variants use slightly less; S and H about 1.5 times more. To this should be added interface circuits, LEDs, etc. It is advisable to choose a supply with considerable conservatism; at least 50 per cent additional capacity is recommended.

Home-made supplies are best based on the 7805 regulators with circuits similar to fig. 13.35. Supplies of up to 5 amps can be constructed in this manner; beyond this commercial supplies should be used. ECL supplies should be based on the 7905 negative regulator.

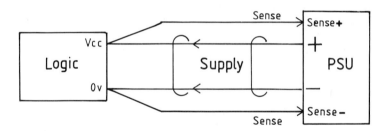

Fig. 13.37 Remote sensing of supply volts.

The need for sensible earthing was outlined in section 13.3. Equally important is the distribution of supply rails in large current systems. Voltage drops down leads can easily exceed the 0.25 volt TTL limit. Large current power supplies with substantial cables are a very real fire hazard. It is essential that overcurrent protection is provided. Commercial supplies usually have remote sensing inputs as in fig. 13.37 so the correct voltage can be maintained at the end of a long supply cable.

Large TTL systems should be protected against overvoltage caused by a supply failure. A common circuit, the so-called crowbar trip, is shown in fig. 13.38. If the supply starts to rise above 6 V, the zener and thyristor turn on and blow the fuse. Commercial supplies usually incorporate some form of overvoltage protection.

Conventional regulator power supplies are bulky, expensive and produce a lot of heat. A recent improvement is the switched mode power supply, which although still expensive gives high current outputs in small, cool packages. This principle is shown in fig. 13.39. The regulator transistor acts as a on/off switch, and feeds the load through a choke filter. The switching is controlled by a fixed-frequency oscillator with mark/space ratio controlled by the error amplifier.

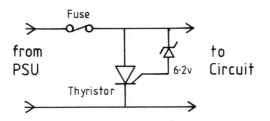

Fig. 13.38 Crowbar overvoltage protection.

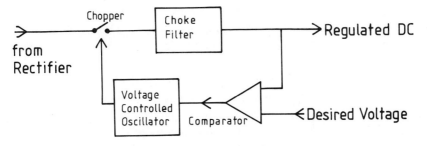

Fig. 13.39 Switching regulator.

The regulating transistor is either off (no current) or full on (low voltage drop) and dissipates little power in either state. The supply therefore runs very cool.

CHAPTER 14
Device Information

14.1 Introduction

This section provides 'snapshot' data sheets for common digital ICs. The data takes the form of a snapshot picture rather than a full rigorous specification. The intention is to provide the user with essential operating information such as sense of clock inputs, device peculiarities, non-obvious restrictions and so on. Full specifications would, in any case, be impossibly long. A full specification for a simple logic gate would occupy some six pages. The detail provided should, however, be adequate for the designer who is not working near a device's limit.

The snapshots cover about 75 per cent of the currently available TTL and CMOS ICs. The choice was based on common usage and, above all, availability. All the devices given below are readily available from many suppliers. Also omitted were devices which could not be readily described in snapshot form such as the 4753 Universal Timer or the 74181/182 Arithmetic Unit which would all require several pages to describe even in brief form. It was also decided to exclude microprocessor-related ROMs and RAMs.

With the above exceptions, the reader should find the data sheets and pin connection diagrams for all devices that he needs. It is strongly suggested that choice of ICs for new designs be made from the selected devices as all are commonly available from many suppliers. There is often a problem obtaining reliable supplies of more esoteric ICs.

14.2 TTL devices

There are five commonly available TTL families (AS, ALS and similar variations are only available for some devices at the time of writing). These have the typical characteristics:

Family	Low o/p sink current (mA)	Gate delay (ns)	MSI delay (ns)	Counter shift register speed (MHz)
L	4	30	60	5
N	16	10	25	20
H	20	6	15	40
LS	8	10	20	30
S	20	3	10	70

These figures are typical of most devices, but the full manufacturer's specification should be checked if a device is to be operated near these limits.

344

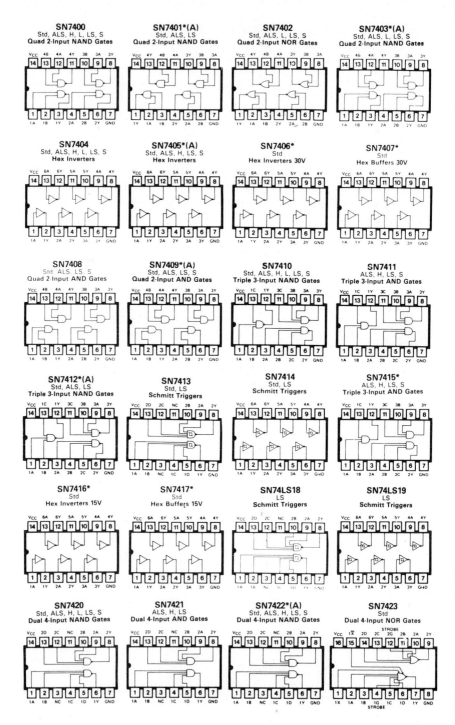

7400 Quad two-input NAND gate
Probably the most popular digital IC.

7401 Quad two-input NAND gate open collector
Voltage rating of output: 5.5 V. Current rating of output: N 16 mA,
LS 8 mA, S 20 mA. Note that pin layout is *not* compatible with
7400.

7402 Quad two-input NOR gate

7403 Quad two-input NAND gate open collector
Voltage rating of output 5.5 V. Current rating of output: N 16 mA,
L 3 mA, S 20 mA. Pin layout *is* compatible with 7400.

7404 Hex inverter

7405 Hex inverter open collector
Pin layout as 7404. Voltage rating of output: 5.5 V. Current rating
of output: N 16 mA, H 20 mA, LS 8 mA, S 20 mA.

7406 Hex inverter open collector
Pin layout as 7404. Voltage rating of output: 30 V. Current rating of
output: 40 mA.

7407 Hex buffer open collector
Pin layout as 7404 but non-inverting. Voltage rating of output 30 V.
Current rating of output 40 mA.

7408 Quad two-input AND gate
Pin layout as 7400 but non-inverting.

7409 Quad two-input AND gate open collector
Pin layout as 7408. Voltage rating of output 5.5 V. Current rating of
output: N 16 mA, LS 8 mA, S 20 mA.

7410 Triple three-input NAND gate

7411 Triple three-input AND gate
Pin layout as 7410 but non-inverting.

7412 Triple three-input NAND gate open collector
Pin layout as 7410. Voltage rating of output 5.5 V. Current rating of output: N 16 mA, LS 8 mA.

7413 Dual four-input NAND Schmitt
Pin layout as 7420. UTP 1.7 volts. LTP 0.9 volts. Input impedance 5K in both input states. Note pins 3, 11 are not used.

7414 Hex Schmitt inverter
Pin layout as 7404. UTP 1.7 volts. LTP 0.9 volts. Input impedance 5K in both states.

7415 Triple three-input AND open collector
Pin layout as 7411. Voltage rating of output 5.5 V. Current rating of output: H 20 mA, LS 8 mA, S, 20 mA.

7416 Hex inverter open collector
Pinning as 7404. Voltage rating of output 15 V. Current rating of output 40 mA.

7417 Hex buffer open collector
Pinning as 7407. Voltage rating of output 15 V. Current rating of output 40 mA.

7418 Dual four-input Schmitt trigger

7419 Hex Schmitt trigger

7420 Dual four-input NAND gate

7421 Dual four-input AND gate
Pinning as 7420 but non-inverting output.

7422 Dual four-input NAND open collector
Pinning as 7420. Voltage rating of output 5.5 V. Current rating of output: N 16 mA, H 20 mA, LS 8 mA, S 20 mA. Pins 11, 3 not used.

7423 Expandable dual four-input NOR 16 pin
Expand with 7460. Output is not tristate, strobe gates output. Expansion inputs do not have full noise immunity.

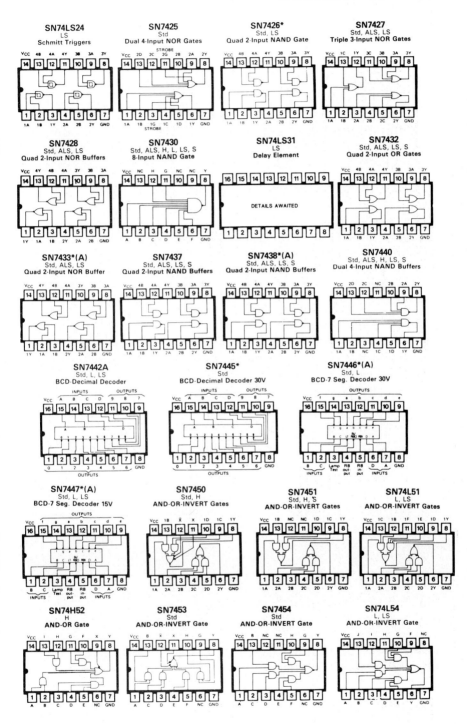

7424 Quad two-input Schmitt trigger

7425 Dual four-input NOR gate with strobe
Strobe gates output. Not tristate.

7426 Quad two-input NAND open collector
Pinning as 7400. Voltage rating of output 15 V. Current rating of output: N 16 mA, LS 8 mA.

7427 Triple three-input NOR gate

7428 Quad two-input NOR buffer
Pinning as 7402. Normalised fan-out 30.

7429
Not used.

7430 Single eight-input NAND gate
Standard supply pins. Pin 13 not used.

7431 Delay element
Not yet released.

7432 Quad two-input OR gate

7433 Quad two-input NOR open collector
Pinning as 7402. Voltage rating of output 5.5 V. Current rating of output: N 48 mA, LS 24 mA.

7435-7436
Not used.

7437 Quad two-input NAND buffer
Pinning as 7400. Normalised fan-out 30.

7438 Quad two-input NAND open collector
Pinning as 7400. Voltage rating of output 5.5 V. Current rating of output: N 48 mA, LS 24 mA, S 60 mA.

7439
Not used.

7440 Dual four-input NAND buffer
Pinning as 7420. Normalised fan-out 30. Pins 3, 11 not used.

7441
Not used (replaced by 74141).

7442 BCD to decimal decoder 16 pin
True BCD input, low logic output (NOT open collector). All ouputs high for invalid input condition.

7443 XS3 to decimal decoder 16 pin
Pinning as 7442. True XS3 input, low logic output (NOT open collector). All outputs high for invalid input condition. Available in N, L.

7444 XS3 Gray to decimal decoder 16 pin
Pinning as 7442. True XS3 Gray input, low logic output (NOT open collector). All outputs high for invalid input condition. Available in N, L.

7445 BCD to decimal decoder/driver 16 pin
Pinning as 7442. True BCD input, low open collector output. Voltage rating of output 30 V. Current rating of output 80 mA.

7446 BCD to seven segment decoder/driver 16 pin
True BCD input open collector. Output low for segment on. Voltage rating of output 30 V. Current rating of output: N 40 mA, L 20 mA. Lamp test input low turns all segments on. Note all possible inputs give some output display. RB inputs/outputs used for ripple blanking (leading/trailing zero suppression). When RB1 is low, a zero will be blanked, other characters displayed. RB0 is an open collector output, goes low when BCD input is 0000 and RB1 is low. If RB0 is taken low display will be off for all inputs. Normally RB0 is connected to RB1 of next most significant stage.

7447 BCD to seven segment decoder/driver 16 pin
As 7446 in all respects but output rating 15 volts. LS current rating 24 mA.

7448 BCD to seven segment decoder 16 pin
As 7446 in operation and pinning. Open collector outputs but internal 2K pull ups to Vcc. True outputs (high for segment on).

Designed for logic uses only, output must not go above Vcc. Normal fan-out. Available in N, LS.

7449 BCD to seven segment decoder
True BCD input. Open collector output OFF for segment ON. (i.e. true outputs). Designed for logic use only, output must not go above Vcc. Normal fan-out. Blanking input low turns all segments off. No lamp test or ripple blanking facilities. Normal supply pins. Available in N, LS. 74249 is improved version. Pinning at end of section.

7450 Dual two-wide two-input AND/OR/INVERT gate
One gate expandable with 7460. Expansion inputs do not have full noise immunity.

7451 AND-OR-INVERT gates
Note pinning differences between H/N, LS version.

7452 Expandable AND-OR gate
H only.

7453 Expandable four-wide AND-OR gate
N version expand with 7460 H version expand with 74H60 or 74H62. Note pinning differences between H and N version.

7454 Four-wide AND-OR-INVERT gates
Available in N, H, L, LS versions with different pinnings. No connections should be made to pins 11, 12 on N, H versions.

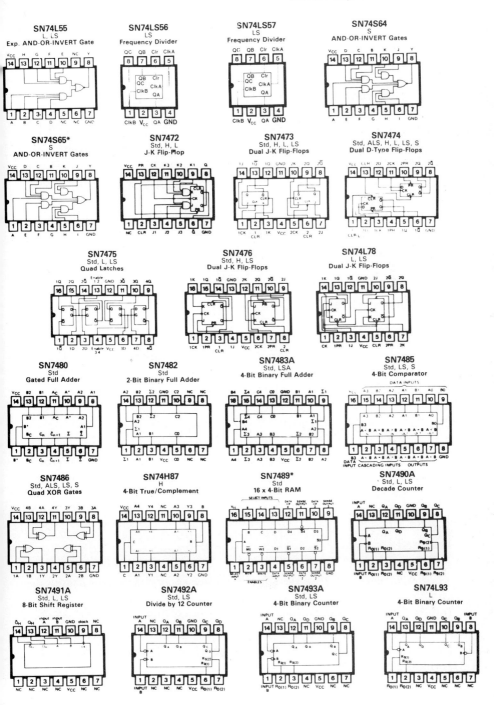

352

7455 Two-wide four-input AND-OR-INVERT gates
H version expandable with 74H60 or 74H62. Available in H, L, LS versions with different pinnings.

7456 Frequency divider

7457 Frequency divider

7460 Dual four-input expanders
For use with 7423, 7450, 7453, 7455. Does not have full noise immunity. Pinning at end of section.

7461 Triple three-input expander
H only (for use with 7452). Pinning at end of section.

7462 Four-wide AND-OR expander
Not commonly available.

7463 Hex current sensing gates
Interfaces to logic elements which source, but not sink, current (e.g. PLAs). Input current $> 200\ \mu A$. Output high. Input current $< 50\ \mu A$ output low. Available only in LS.

7464 4-2-3-2 AND-OR-INVERT gate

7465 4-2-3-2 AND-OR-INVERT gate
As 7464 with open collector output. For logic use only; output must not go above Vcc.

7466-7469
Not used.

7470 AND gates JK positive edge triggered JK flip-flop with preset and clear
Inputs $\bar{J}$ and $\bar{K}$ must be held low if not used. Preset and clear inputs are active low, and should only be used when clock is low. Clock is positive edge triggered; outputs change when clock goes from 0 to 1. Available only in N.

74H71 AND-OR gated JK master slave flip-flop with preset
Not commonly used.

74L71 AND gated RS master slave flip-flop with preset/clear
Not commonly used.

7472 AND gated JK master slave flip-flop with preset and clear
Preset and clear are active low and override clocked inputs. Outputs change when clock goes from 1 to 0. Level triggered device; inputs should not change when clock high.

7473 Dual JK flip-flop with clear
Clear active low override clocked inputs. N, H, L are level triggered devices and inputs should not change whilst clock is high. LS version is edge triggered. On all versions outputs change when clock goes from 1 to 0.

7474 Dual D-type flip-flop with preset and clear
Presets and clear are active low and override clocked inputs. All devices are edge triggered, and outputs change when clock goes from 0 to 1.

7475 Quad latch 16 pin
Four hold/follow latches grouped into two sets of two. Hold follow controlled by G: with G = 1, output follows D input; with G = 0 output holds. Not true clocked D type. Non standard supplies.

7476 Dual JK with bistable and clear
Presets and clear are active low and override clock. N and H are level triggered devices and inputs should not change whilst clock is high. LS version is edge triggered. On all versions outputs change when clocked from 1 to 0. Non standard supplies.

7477 Quad latch
Functionally as 7475 in 14 pin IC. $\bar{Q}$ outputs not available. Non standard supplies. Not available in DIL package.

7478 Dual JK flip-flop with preset, common clear and common clock
Preset, clear active low and override clocks inputs. H and L versions are level triggered, and inputs should not change when clock is high. LS version is edge triggered. Outputs change when clock goes from 1 to 0.

7479
Not used.

7480 Gated full adder

7481 16 bit RAM

7482 Two-bit full adder
C0 is carry in, C2 carry out. Internal look ahead carry. Non standard supplies.

7483 Four-bit full adder
C0 is carry in, C4 carry out. Internal look ahead carry. Non standard supplies.

7484 16 bit RAM

7485 Four-bit magnitude comparator
Compares two four-bit words A, B and produces true outputs A$<$B, A=B, A$>$B. Cascade inputs are used to compare words of greater than 4 bits, cascading from Least Significant to More Significant, outputs being taken from most significant device. Cascade inputs on least significant device should be wired A = B High, A $<$ B and A $>$ B Low. Available in N, LS, S (and in L with different connections).

7486 Quad exclusive OR

7487 Four-bit true/complement/zero/one element
Output state controlled by BC inputs

B	C	Output
0	0	complement
0	1	true
1	0	All high
1	1	All low

Available in H (true/complement switching can also be provided by 7486).

7488 256 bit ROM

7489 64 bit RAM

7490 Four-bit non synchronous decade counter

Preset 9 and clear AND gated inputs are active high, if not used must be held low. Independent divide by two and divide by five counters. In normal BCD use, clock input to pin 14 and link pin 12 to pin 1. If divide by five precedes divide by two symmetrical divide by ten output is obtained. Counter steps on negative clock edge. Cascadable to form multistage ripple counter. Non standard supply connection.

7491 8 bit shift register

Serial in, serial out in true/complement form. Non standard supply connection.

7492 Divide by twelve non synchronous counter

Reset AND gated inputs are active high, if not used must be held low. Independent divide by two and divide by six counters. In normal use, clock input to pin 14 and link pin 12 to pin 1. If divide by six precedes divide by two, symmetrical divide by twelve output is obtained. Counter steps on negative clock edge. Cascadable to form multistage ripple counter. Non standard supply connections.

7493 Four-bit binary non synchronous counter

Clear AND gated inputs are active high, if not used must be held low. Independent divide by two and divide by eight counters. Counter steps on negative clock edge. Cascadable to form multistage ripple counter. Non standard supply connections.

356

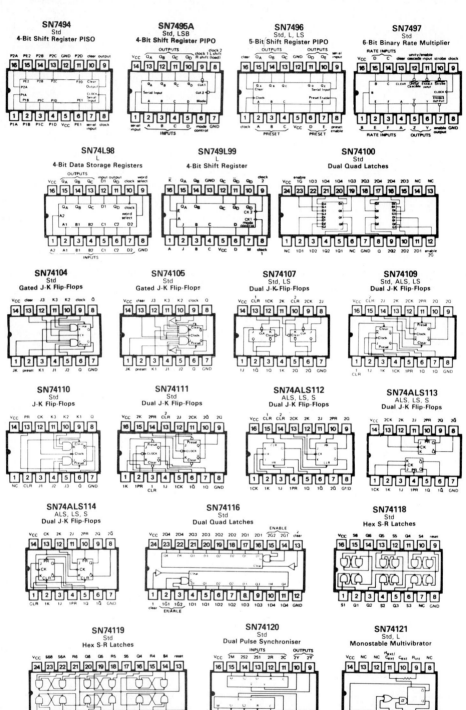

7494 Four-bit shift register
Two preset inputs denoted 1, 2 selected by preset enables 1, 2 active high. Presets are asynchronous. Common clear active high. Shift on positive edge of clock. Non standard supply connections.

7495 Four-bit shift register
Bidirectional operation. Negative edge on SRT shifts one place right (defined as A-B-C-D). Negative edge on SLT shifts one place left. With mode control low, register can shift. With mode control high, data on parallel load inputs is loaded into register when a low-high-low pulse is applied to the SLT input. Note SLT performs two functions. Available in N, LS (and L with different pinning).

7496 Five-bit shift register
Clear is active low. Preset active high. Shifts on positive edge of clock. Non standard supplies.

7497 Six-bit binary rate multiplier

7498 Four-bit data selector/storage register

7499 Four-bit bidirectional shift register

74100 Eight-bit hold follow latch 24 pin IC
Enable high to follow, low to hold.

74101 AND-OR gated JK flip-flop
Available in H. Not commonly used.

74102 AND gated JK flip-flop
Available in H. Not commonly used.

74103 Dual JK flip-flop
Available in H. Not commonly used.

74104 Gated JK flip-flop

74105 Gated JK flip-flop

74106 Dual JK flip-flop
Available in H. Not commonly used.

74107 Dual JK flip-flop
Clear active low. Outputs change when clock goes from 1 to 0. Level triggered; inputs should not change whilst clock is high.

74108 Dual JK flip-flop
Available in H. Not commonly used.

74109 Dual J$\overline{K}$ positive edge triggered flip-flop 16 pin
Preset and clear active low, override clocked inputs. Positive edge triggered, outputs change when clock goes from 1 to 0. Note $\overline{K}$ input.

74110 AND gated master slave JK flip-flop

74111 Dual JK master slave flip-flop

74112 Dual JK negative edge triggered flip-flop 16 pin
Preset and clear active low, override clocked inputs. Negative edge triggered, outputs change when clock goes from 1 to 0.

74113 Dual JK negative edge triggered flip-flop
Preset active low, overrides clocked input. Negative edge triggered, outputs change when clock goes from 1 to 0.

74114 Dual JK negative edge triggered flip-flop
Note common clock and common clear. Preset and clear active low, override clocked inputs. Negative edge triggered, outputs change when clock goes from 1 to 0.

74115
Not used.

74116 Dual four-bit hold follow latch 24 pin
Follows when both enables are low. Clear active low overrides hold/follow function.

74117
Not used.

74118 Hex SR latch

74119 Hex SR latch

74120 Dual pulse synchroniser 16 pin

74121 Non re-triggerable monostable
A inputs active low, B inputs active high. All inputs edge triggered. If B input is used, one A input should be held low. B input is Schmitt trigger. Timing capacitor between pins 11, 10. Electrolytics positive to pin 11. Timing resistor from pin 11 to Vcc. Internal 2K (L 4K) resistor to pin 9 may be used instead. Period 0.7 RC. Minimum R is 1K5, maximum advised C 10 μF.

360

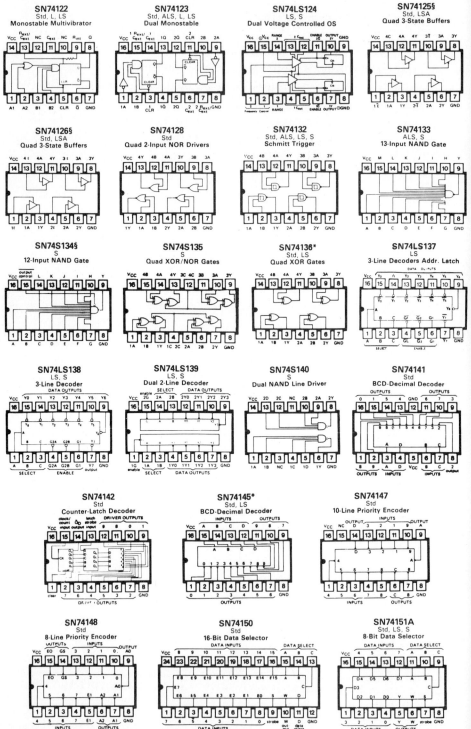

74122 Retriggerable monostable

A inputs active low, B inputs active high. If B inputs are used, at least one A input must be held low. Edge triggered and re-triggerable. Clear is active low and inhibits inputs. If input conditions are true, removal of clear will trigger device. Timing capacitor between pins 13 and 11 (electrolytic positive to pin 13). Timing resistor from pin 13 to Vcc. Internal resistor (10K N and LS 20K L) available on pin 9. Period 0.4 RC. Minimum R 5K. Maximum advised C 10 μF. Use diode to prevent reverse voltage on capacitor with N and L versions. With LS version connect C ext terminal to 0 V when using electrolytic.

74123 Dual re-triggerable monostable

Comments as 74122 except no internal resistor.

74124 Dual VCO

74125 Quad tristate buffer

Output is OFF (high impedance) when control input is HIGH.

74126 Quad tristate buffer

Output is OFF (high impedance) when control input is LOW.

74127

Not used.

74128 Quad line driver

NOR line driver, pinning as 7402. Capable of driving terminated 75 ohm and 50 ohm lines.

74129-74131

Not used.

74132 Quad two-input NAND Schmitt trigger

Pinning as 7400.

74133 13 input NAND gate

74134 12 input NAND gate with tristate output

Output is OFF (high impedance) when control input is HIGH.

74135 Quad XOR/NOR gate
$Y = (A \oplus B) \oplus C.$

74136 Quad XOR gate

74137 3 line decoder address latch

74138 3 to 8 line decoder/demultiplexer 16 pin
Decodes one of eight outputs (active low) on state of ABC inputs. True and complement enable; outputs enabled when G1 = 1 and both G2 = 0. True and complement enables allows decoding of 24 lines with three devices.

74139 Dual 2 to 4 line decoder/demultiplexer 16 pin
Decodes one of four outputs (active low) on state of AB inputs. Enable active low.

74140 Dual four-input NAND line driver
Capable of driving 75 ohm and 50 ohm lines. Pinning as 7420.

74141 BCD to decimal decoder driver
High voltage open collector outputs for driving cold cathode tubes. Outputs rated at 60 volts 7 mA. Non BCD inputs blank all outputs.

74142 Counter/latch/nixie driver
IC incorporates BCD counter/latch and one of ten decoder/driver with outputs rated at 60 volts 7 mA. Counter steps on positive clock edge. $\bar{Q}_D$ output is for cascading and goes low during outputs 8, 9. Clear is active low and inhibits counting. Hold/follow latch controlled by latch strobe; low for follow, high for hold. Available in N.

74143 Counter latch decoder seven segment 15 mA, 5 V outputs plus latch output
Available in N. Pinning at end of section.

74144 Counter latch decoder seven segment 25 mA, 15 V outputs plus latch output
Available in N. Pinning at end of section.

74145 BCD to decimal decoder/driver 16 pin
Pinning as 7442 open collector. Outputs rated at 15 volts, 25 mA.

74146
Not used.

74147 BCD priority encoder 16 pin
Inputs active low. Output BCD shows highest active input.

74148 Octal priority encoder 16 pin
Inputs active low. Output octal shows highest active input. Enable active low. Eo goes low if ANY input is low, Gs goes low if any input is low AND enable is low. Cascadable.

74149
Not used.

74150 1 of 16 data selector 24 pin
Address ABCD selects one of 16 inputs. Enable active low. Output data is COMPLEMENT of selected data.

74151 1 of 8 data selector 16 pin
Address ABC selects one of 8 inputs. Enable active low. Output data available in time and complement form.

364

74152 1 of 8 data selector
Available only in 54 series.

74153 Dual 1 of 4 data selector 16 pin
Common address lines. AB address selects one of 4 inputs. Independent enables (strobes) active low. True output.

74154 4 bit decoder/demultiplexer 24 pin
4 bit binary to 16 line output decoder/demultiplexer. Enable is active low. Selected output follows data input; other outputs are high. For use as simple decoder pins 19 and 18 should be held low.

74155 Decoder/demultiplexer
As 74156 with logic outputs.

74156 Decoder/demultiplexer
Open collector outputs rated 5 volts at 16 mA. Common AB address selects one of 4 inputs. Enable (strobe) is active low. Data IC appears at selected 1 output. Data 2C appears inverted at selected 2 output. Unselected outputs are high. Two parts of IC can be used to form a 3 to 8 line decoder or a 1 to 8 demultiplexer.

74157 Quad 2 to 1 line data selector/multiplexer
Two four bit data words A, B are applied to the IC. 4 bit output, Y, follows A input if select input is low; Y follows B input if select is high. Strobe low for normal operation, strobe high Y outputs all low.

74158 Quad 2 to 1 line data selector/multiplexer
As 74157 with inverted output.

74159 4 to 16 line decoder/demultiplexer 24 pin
As 74154 with open collector outputs. Outputs rated 5 volts at 16 mA.

74160, 74161 Synchronous four-bit counters, direct clear
74160 BCD, 74161 binary. Otherwise identical. Clear active low, overrides clock. For normal counting P, T and load, inputs are high. Counter advances on 0 to 1 edge of clock input. To load the counter, the required data is placed on the load inputs, and the load terminal taken low. The data is loaded at the next clock pulse. P and T are enable inputs, both must be high for counting to occur. For multi-

stage counters, carry out should be connected to T input of next more significant IC. Clock should be common to all ICs.

74162, 74163 Synchronous four-bit counter, synchronous clear
74162 BCD, 74163 binary. As 74160/1 except that clear is synchronous, i.e. occurs on positive clock edge after clear input taken low. Clear overrides counting. This feature is useful for modifying count sequence.

74164 Eight-bit serial in parallel out shift register
Eight-bit single direction (A to H) shift register. Data shifted in for each positive clock pulse is the AND function of S1, S2 (i.e. both S1 and S2 must be high to shift a 1 into A). Clear active low overrides clock.

74165 Parallel input serial output eight-bit shift register
Eight-bit single direction (A to H) shift register. Parallel loading achieved by placing data on parallel inputs and taking "load" input low. Shifting occurs on positive clock edge. Clock enable is low to permit shifting. Serial input shifts into A. The Q, Q̄ outputs of stage H only are available.

74166 Parallel input serial output eight-bit shift register
Eight-bit single direction (A to H) shift register. Synchronous parallel load achieved by placing data on inputs and taking "load" input low prior to positive clock edge. Shifting occurs on positive clock edge. Clear active low overrides clock.

74167 Decade rate multiplier

74168 Decade synchronous up/down counter
All operations synchronous on positive clock edge. Normal counting P, T inputs held low. Direction line high for count up, low for count down. P or T high inhibits all operations. Normally carry out connected to T input for multistage counting. Synchronous load achieved by placing data on inputs and taking load input low prior to positive clock edge.

74169 Binary synchronous up/down counter
As 74168 but binary count.

74170 4 × 4 register file

74171 Quad D-type flip-flops

74172 16-bit register file

368

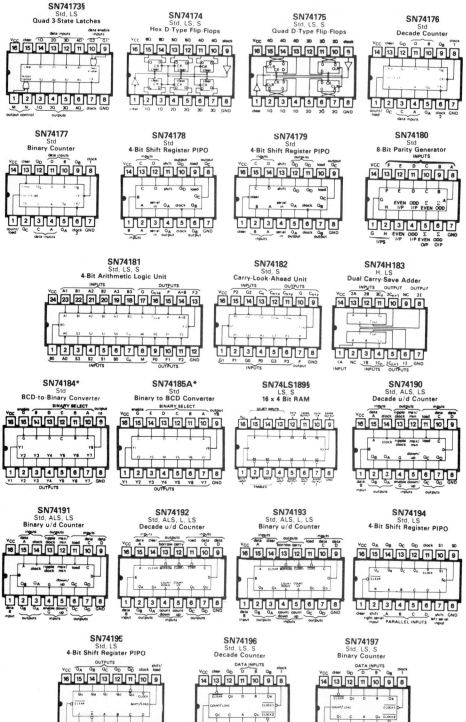

74173 Quad D-type, tristate output

True D types. Data loaded on positive clock edge. Clear active high overrides clock. Both data enable lines must be low to load data, if either is high output states are unchanged by positive clock edge. Both output control lines must be low to enable outputs, otherwise outputs are in high impedance state. Output control lines do NOT affect data loading.

74174 Hex D-type

True D types. Clear active low overrides clock. Data loaded on positive clock edge.

74175 Hex D-type

Operation as 74174.

74176 Presettable decade counter/latch

Pinning and function as 74196.

74177 Presettable binary counter/latch

Primary and function as 74197.

74178 Four-bit shift register

74179 Four-bit shift register

74180 Nine-bit parity generator

Generates odd/even parity outputs for 9 bit input word A-H. Carry inputs used to cascade ICs. If both are low, both outputs are high regardless of A-H. If both carry inputs are low, both outputs are low regardless of A-H. Other combinations are best summarised by a table:

Number of 1s	Even i/p	Odd i/p	Even o/p	Odd o/p
Even	H	L	H	L
Odd	H	L	L	H
Even	L	H	L	H
Odd	L	H	H	L

74181 Arithmetic and logic unit

Performs 16 arithmetic and logic operations on two four-bit words. This is a complex device whose description is beyond the scope of a snapshot description.

74182 Look ahead carry generator
For use with 74181.

74183 Dual adder

74184 BCD to binary converter
See section 8.10.

74185 Binary to BCD converter
See section 8.10. Pinning as 74184.

74186 512 bit PROM

74187 1024 bit ROM

74188 256 bit PROM

74189 64 bit RAM

74190 Synchronous up/down BCD counter
Synchronous counting, asynchronous preset. Normal counting on positive clock edge with enable low, and load high. If direction line is low counter counts UP. If direction line is high counter counts DOWN. To load data, load input is taken low. Load overrides clock. ICs can be cascaded by linking ripple clock output to enable input of the succeeding counter if parallel clocking is used, or by enabling is used.

74191 Synchronous up/down binary counter
As 74190 with binary count.

74192 Synchronous up/down BCD counter
Independent count up/down clocks. Counter steps on positive edge of clock. Other clock must be high. Clear input active HIGH, overrides count inputs. Load active low overrides clocked inputs. For cascading link carryout to succeeding count up clock, borrow out to succeeding count down clock.

74193 Synchronous up/down binary counter
As 74192 with binary count.

74194 Four-bit bidirectional shift register
Shift right defined as A to D. Synchronous load/shift. Asynchronous clear active low overrides clock. Mode determined by S0, S1 inputs:

S1	S0	Action on positive clock edge
0	0	Do nothing, outputs unchanged
0	1	Shift right
1	0	Shift left
1	1	Parallel load

74195 Four-bit parallel load shift register
Single direction (A to D) shift register. Shifts on positive clock edge. Synchronous load with active low load input. Asynchronous clear overrides clock. Shift in input features $J\overline{K}$ operation as below:

J	$\overline{K}$	QA after clock pulse
0	1	Unchanged
0	0	0
1	1	1
1	0	Toggle

74196 Presettable decade counter
When count/load input is low, outputs follow data inputs regardless of clock. With count/load input high acts as counter. Independent divide by five and divide by two stages. Normal BCD count link QA output to clock 2 input and count input to clock 1. For symmetrical divide by ten, link QD output to clock 1 input and count input to clock 2. Counter steps on negative clock edge. Clear active low overrides clock.

74197 Presettable binary counter
Binary version of 74196 with independent divide by eight and divide by two stages. Pinning as 74196.

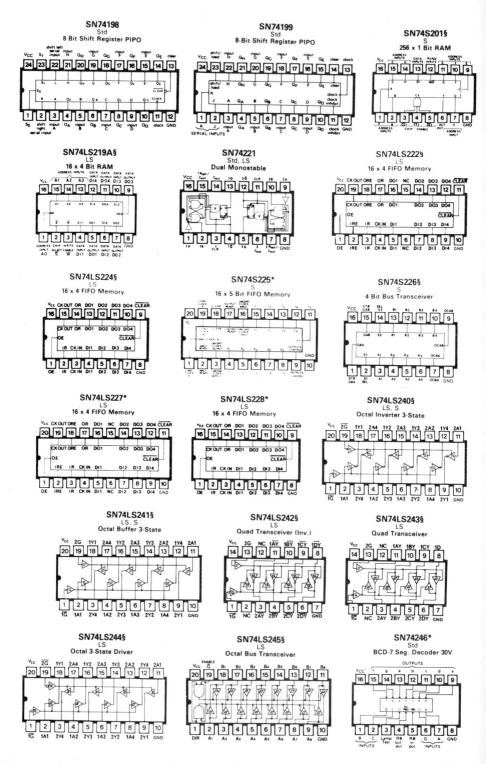

74198 Eight-bit bidirectional shift register

74199 Eight-bit bidirectional shift register

74200 256 bit RAM

74201 256 bit RAM

74202 256 bit RAM

74203-74206
Not used.

74207 256 by 4 bit RAM

74208 256 by 4 bit RAM tristate output

74209-74213
Not used.

74214 1024 bit RAM

74215 1024 bit RAM

74216-74218
Not used.

74219 16 X 4 bit RAM

74220
Not used.

74221 Dual monostable with Schmitt trigger input
Functionally identical to 74123.

74222 16 X 4 FIFO memory

74223
Not used.

74224 16 X 4 FIFO memory

74225 FIFO 16×5 bit memory

74226 Four-bit latched bus transceiver

74227 16 × 4 FIFO memory

74228 16 × 4 FIFO memory

74229-74239
Not used.

74240 Octal tristate inverting buffer
Inverting buffers. Gate signal LOW enables outputs. Can source 15 mA and sink 24 mA (LS) 64 mA (S). Can drive terminated lines down to 135 ohms.

74241 Octal tristate non-inverting buffer
Non-inverting buffer. Pinning as 74240 with no inversion and true gate input on pin 19. Gate signal on pin 19 HIGH enables outputs. Gate signal pin 1 LOW enables outputs. Other details as 74240.

74242 Quad transceiver, inverted output
Note pin 13 true gate, pin 1 complement. Normally these are linked to form direction signal. With this direction signal high, data flows from B to A. With direction signal low, data flows from A to B. Data output is inverted.

74243 Quad transceiver, non-inverted output
Pinning as 74242 but true output data. Other details as 74240.

74244 Octal tristate non-inverting buffer
Non-inverting buffer. Pinning as 74240 with no inversion. Gate signals LOW enable outputs. Other details as 74240.

74245 Octal bus transceiver
Pin 19 low to enable data transfer. Pin 1 high data flow A to B. Pin 1 low data flow from B to A. Data output true.

74246 BCD to seven segment decoder 30 V output
Functionally identical to 7446 with different segment display for 6 and 9.

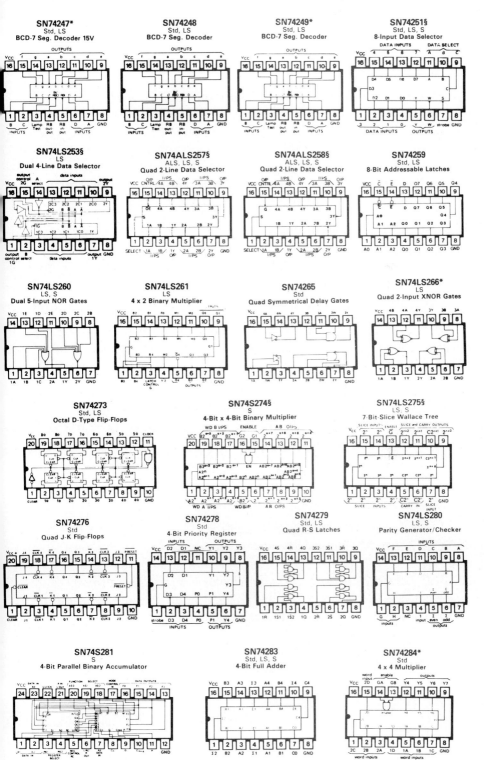

74247 BCD to seven segment decoder 15 V output
Functionally identical to 7447 with different segment display for 6 and 9.

74248 BCD to seven segment decoder true logic output
Functionally identical to 7448 with different segment display for 6 and 9.

74249 BCD to seven segment decoder true output
Pinning as 7448 (improved 7449 with addition of lamp test and ripple blanking). Different segment display for 6 and 9.

74250
Not used.

74251 1 of 8 data selector with three state output
ABC address lines. Selected data available in true/complement form. Enable input LOW to enable outputs.

74252
Not used.

74253 Dual two-input multiplexer with three state outputs
Common AB address lines. Individual enable inputs (LOW to enable outputs). True outputs.

74254, 74255
Not used.

74256 Dual four-bit addressable latch
Available in LS. Pinning at end of section.

74257 Quad 2 line to 1 line data selector with tristate output
Select line low A input selected; high B input selected. True output data. Output control LOW to enable outputs.

74258 Quad 2 line to 1 line data selector with tristate output
As 74257 but complement output data.

74259 Eight-bit addressable latch
ABC address lines select latch. Input data applied to Din. Operation controlled by clear and enable inputs as below:

Clear	Enable	Function
H	H	Load addressed latch, other latch outputs unaffected
H	H	Latch outputs unaffected (quiescent state)
L	L	Addressed latch output enables, other outputs low
L	H	All latches cleared

To avoid transients, enable and clear should be held high whilst changing addresses to avoid glitches causing problems. Direct replacement for Fairchild 9334 and TIM 9906.

74260 Dual five-input NOR

74261 2 × 4 bit multiplexer

74262-74264
Not used.

74265 Quad true/complement buffers
True/complement outputs appear simultaneously with no skew.

74266 Quad two-input Exclusive NOR gates with open collector output
For use with logic systems, outputs rated at 5.5 V.

74267-74269
Not used.

74270 2K ROM

74271 2K ROM

74272
Not used.

74273 Octal D-type
Common clock. Outputs change on positive clock edge. Clear active low overrides clocked inputs.

74274, 74275 Binary multipliers
See section 8.9.

74276 Quad J$\bar{K}$ flip-flop

74277
Not used.

74278 Four-bit priority register

74279 Quad SR flip-flop

74280 Nine-bit parity generator
Up graded 74180.

74281 Four-bit arithmetic unit
Performs 15 functions including arithmetic, logic, shifts.

74282
Not used.

74283 Four-bit binary adder
Functionally identical to 7483 but with conventional supply pinning.

74284, 74285 Binary multipliers
See section 8.9.

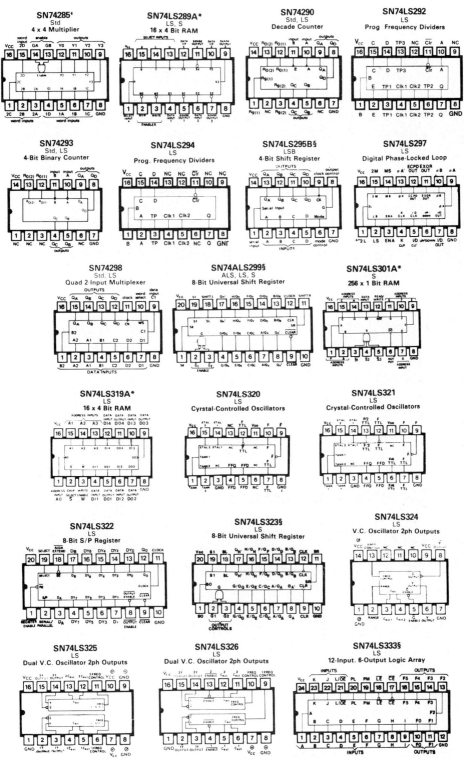

74286
Not used.

74287 1K PROM

74288 256 bit PROM

74289 64 bit RAM

74290 Decade counter
Functionally identical to 7490 with conventional supply pinning.

74291
Not used at present.

74292 Programmable frequency divider

74293 Four-bit binary counter
Functionally identical to 7493 with conventional supply pinning.

74294 Programmable frequency divider

74295 Four-bit bidirectional shift register
Available in LS.

74296
Not used.

74297 Digital phase-locked loop

74298 Quad two-input multiplexer with storage
Word select input selects one of two four-bit input words; low word 1, high word 2. Data stored on negative clock edge.

74299 Eight-bit shift register with tristate outputs
Input and output share common pins. Operation controlled by two select lines S0, S1 as below. Clocked functions on positive edge.

S1	S0	Function
0	0	Quiescent state. Outputs unchanged
0	1	Shift right (A to H) on clock pulse. Shift Rt i/p to A

1   0   Shift left (H to A) on clock pulse. Shift Lt i/p to H.
1   1   Parallel load. I/O lines go to high impedance state and
        act as inputs

Enable inputs G1, G2 control output state (except when loading)
BOTH must be LOW to enable outputs. Clear active low overrides
clocked functions. Available in LS, S.

74300 256 bit RAM

74301 256 bit RAM

74302 256 bit RAM

74303-74313
Not used.

74314/315 1K RAM

74316-74318
Not used.

74319 16 X 4 bit RAM

74320 Crystal oscillator

74321 Crystal oscillator

74322 Eight-bit shift register

74323 Eight-bit shift register with tristate outputs
As 74299 with synchronous clear active high.

74324 Voltage controlled oscillator

74325, 74326, 74327 Dual voltage controlled oscillator

74328-74331
Not used.

74333 12 in 6 out logic array

382

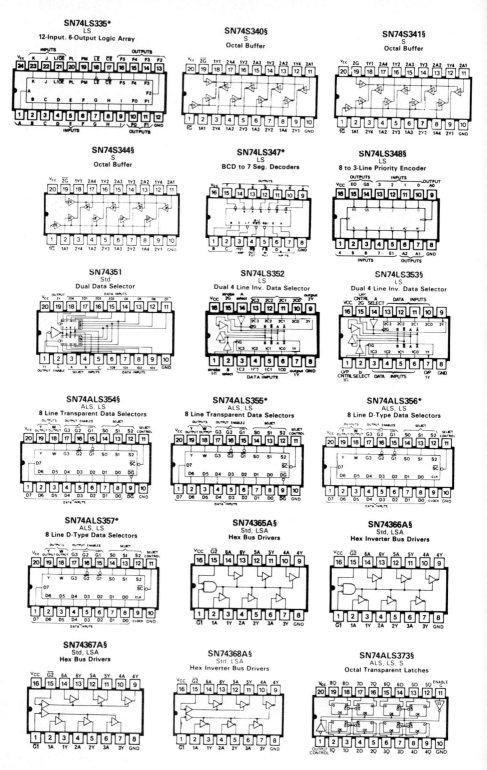

74334
Not commonly used.

74335 12 in 6 out logic array

74336-74339
Not commonly used.

74340/341/344 Octal buffer

74345/346
Not used.

74347 BCD to seven segment decoder
Pinning as 7446/7447

74348 Octal priority encoder with tristate output
(TIM 9908)

74349, 74350
Not used.

74351 Dual data selector

74352 Dual 1 of 4 data selector
Pinning and details as 74153 with INVERTED outputs.

74353 Dual two-input multiplexer with three state output
Pinning and details as 74253 with INVERTED outputs.

74354 8 line transparent data selector

74355 8 line transparent data selector

74356 8 line D-type data selector

74357 8 line D-type data selector

74358-361
Not used.

74362 Clock generator for TMS 9900 microprocessor

74363 Octal hold/follow latch tristate output
NOT true D-type. Enable high output follows input, enable low output holds. Output control low to enable outputs. Schmitt trigger enable line. High output voltage (3.65 V min) so can interface directly to CMOS. Pinning as 74373.

74364 Octal D-type tristate output
True D type. Pinning as 74373. Positive edge triggered clock.

74365 Hex tristate non-inverting buffer
Both enables low to enable outputs.

74366 Hex tristate inverting buffer
As 74365 with inverted data outputs.

74367 Hex tristate non-inverting buffer
Separate 4 line and 2 line sections. Enables low to enable outputs.

74368 Hex tristate inverting buffer
As 74367 but inverted data outputs.

74369
Not used.

74370 2K ROM

74371 2K ROM

74372
Not used.

74373 Octal hold/follow latch with tristate output
As 74363 with normal TTL output voltage levels.

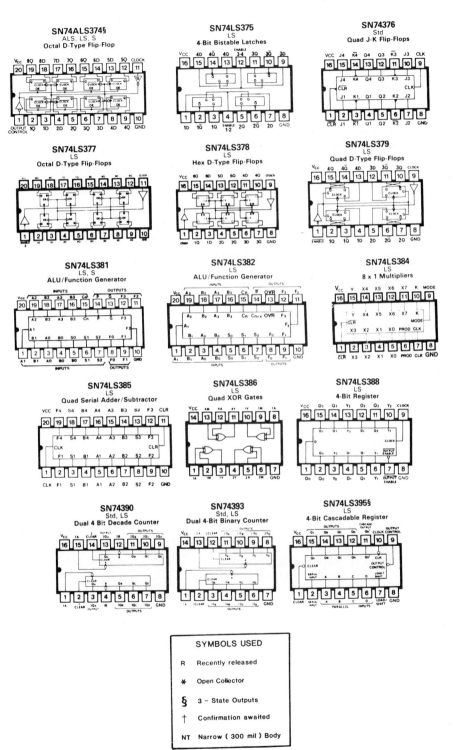

SN74ALS374§
ALS, LS, S
Octal D-Type Flip-Flop

SN74LS375
LS
4-Bit Bistable Latches

SN74376
Std
Quad J-K Flip-Flops

SN74LS377
LS
Octal D-Type Flip-Flops

SN74LS378
LS
Hex D-Type Flip-Flops

SN74LS379
LS
Quad D-Type Flip-Flops

SN74LS381
LS, S
ALU/Function Generator

SN74LS382
LS
ALU/Function Generator

SN74LS384
LS
8 x 1 Multipliers

SN74LS385
LS
Quad Serial Adder/Subtractor

SN74LS386
LS
Quad XOR Gates

SN74LS388
LS
4-Bit Register

SN74390
Std, LS
Dual 4 Bit Decade Counter

SN74393
Std, LS
Dual 4-Bit Binary Counter

SN74LS395§
LS
4-Bit Cascadable Register

SYMBOLS USED

R   Recently released

*   Open Collector

§   3 – State Outputs

†   Confirmation awaited

NT   Narrow (300 mil) Body

74374 Octal D-type tristate output
As 74364 with normal TTL output voltage levels.

74375 Four-bit hold/follow latch

74376 Quad JK̄ flip-flop

74377 Octal D-type
Similar to 74273, but pin 1 is clock enable and must be low to allow clocking. Positive edge triggered clock.

74378 Hex D-type
Similar to 74174, but pin 1 is clock enable and must be low to allow clocking. Positive edge triggered clock.

74379 Quad D-type
Similar to 74175 but pin 1 is clock enable and must be low to allow clocking. Positive edge triggered clock.

74380
Not used.

74381 Arithmetic unit

74382 ALU/function generator

74383
Not used.

74384 8 × 1 multiplier

74385 Quad serial adder/subtracter

74386 Quad XOR gate
(7486 usual choice).

74387 1K PROM

74388 Four-bit register

74389
Not used.

74390 Dual decade counter
Dual 7490 with divide-by-5 and divide-by-2 stage. Clear active high overrides clock. Counter advances on 1 to 0 clock input. Available in N, LS.

74391, 74392
Not used.

74393 Dual four-bit binary counter
Dual 7493. Clear active high overrides clock. Counter advances on 1 to 0 clock input. Available in N, LS.

74394
Not used.

74395 Four-bit shift register with three-state output
Available in N, LS.

Remaining devices are either unused, special microprocessor-related functions, duplicates of earlier devices or uncommon.

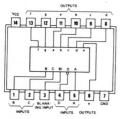

SN5449 (W)
SN54LS49 (J, W)   SN74LS49 (J, N)

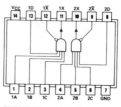

SN5460 (J)   SN7460 (J, N)
SN54H60 (J)   SN74H60 (J, N)

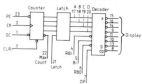

SN54H61 (J)   SN74H61 (J, N)

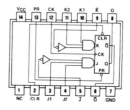

SN5470 (J)   SN7470 (J, N)

74143/4 Counter/Latch/Decoder/Driver

A SELECTION OF POPULAR
TEXAS INSTRUMENTS 75 SERIES

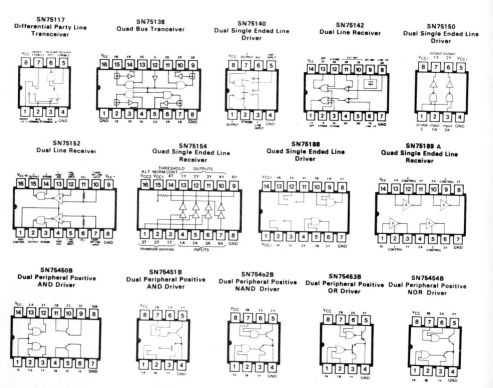

SN75117
Differential Party Line
Transceiver

SN75138
Quad Bus Tranceiver

SN75140
Dual Single Ended Line
Driver

SN75142
Dual Line Receiver

SN75150
Dual Single Ended Line
Driver

SN75152
Dual Line Receiver

SN75154
Quad Single Ended Line
Receiver

SN75188
Quad Single Ended Line
Driver

SN75189 A
Quad Single Ended Line
Receiver

SN75450B
Dual Peripheral Positive
AND Driver

SN75451B
Dual Peripheral Positive
AND Driver

SN75452B
Dual Peripheral Positive
NAND Driver

SN75453B
Dual Peripheral Positive
OR Driver

SN75454B
Dual Peripheral Positive
NOR Driver

14.3 CMOS devices

CMOS devices are generally available in A (original) and B (buffered) versions. B versions should be used for all new designs.

The typical characteristics below can be assumed for most devices, but manufacturers data sheets should be checked if a device is to be used near these figures.

	5 volts	15 volts
Output sink/source current	0.5 mA	2 mA
Gate delay	60 ns	20 ns
MSI delay	250 ns	100 ns
Counter speed	2 MHz	5 MHz
Clock edge speeds	5 μs maximum	

It should be remembered that *all* CMOS inputs must go somewhere, even on unused portions of multi-gate ICs.

CMOS ICs all have positive supplies at the highest pin number and 0 V at the opposite diagonal (e.g. positive pin 16, 0 V pin 8 for a 16 pin IC) unless otherwise stated.

4000 Dual three-input NOR gate plus one inverter

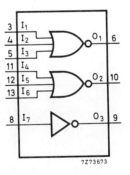

7Z73673

4001 Quad two-input NOR gate

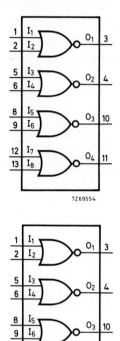

7Z69554

7Z69554

4002 Dual four-input NOR gate

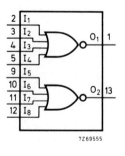

7Z69555

4003-4005
Not used.

4006 Serial-in, serial-out shift register

This IC contains four separate shift registers; two 4-bits long, two 4/5-bits long, which can be connected to form any required length. Shifts occur on negative clock edge. Pins 8 and 11 are outputs only.

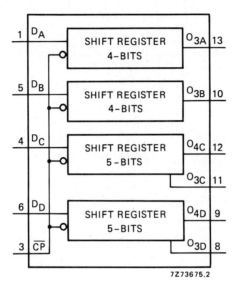

7Z73675.2

4007 General-purpose CMOS array

A not commonly-used package of CMOS transistors which can be used by the designer for a variety of functions such as analog switches, DACs, etc.

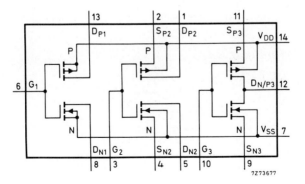

7Z73677

4008 Four-bit full adder
Internal look ahead carry. Carry Out (Co) and Carry In (Cin).

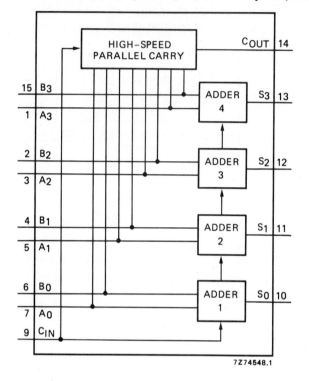

7Z74548.1

4009 Hex inverting buffer
Replaced by improved 4049.

4010 Hex non-inverting buffer
Replaced by improved 4050.

4011 Quad two-input NAND

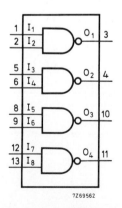

7Z69562

4012 Dual four-bit NAND

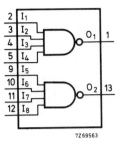

4013 Dual D-type

True D-type. Outputs change on positive clock edge. Set and reset active high overrides clocked input.

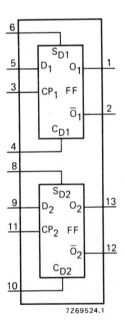

4014 Eight-bit shift register

Single direction (0 to 7). Shifting occurs on positive clock edge if load (PE) is low. With load (PE) high, data on 0 to 7 is synchronously loaded on positive clock edge. Note that although all eight bits can be pre-loaded only outputs 567 can be read. D_S shifts into 0 and is used for cascading.

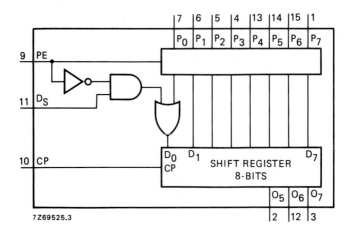

7Z69525.3

4015 Dual four-bit shift register

Two independent serial-in, parallel-out four-bit shift registers. Single direction (0 to 3) with D shifting to O. Shifts on positive clock edge. Reset active high overrides clock.

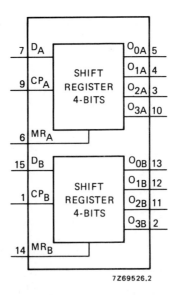

7Z69526.2

4016 *Quad analog switch (transmission gate)*

Switch made (on) when control input high. Switch open (off) when control input low. ON resistance 300 ohms. 4066 has lower ON resistance. See section 12.2.

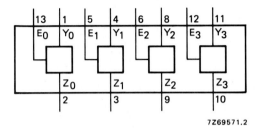

7Z69571.2

4017 *Synchronous decade counter with decoder*

The counter is based on a five-stage Johnson counter with ten decoded active high outputs, and an active low output for cascading. The counter can advance on positive edges on CPO (with CPI low) or negative edges on CPI (with CPO high). The cascading output is a symmetrical output that goes low for counts 5 to 9. Reset is active high, overrides the clocks and puts outputs O and Cascade Out high, all others low.

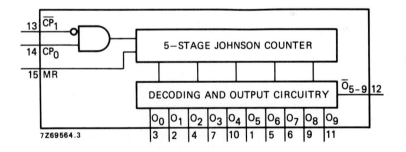

7Z69564.3

4018 Synchronous presettable divide counter

The counter is based on a five-stage Johnson counter, and can be made to divide by any integer between 2 and 10. This is achieved by the five parallel load inputs and the D input to the first stage of the counter. Data is parallel-loaded whilst PL is high, overriding the clock. The count sequence is modified by connecting the D input as below:

Divide by	*D connected to*
10	$\bar{O}_4$
9	$\bar{O}_4$ and $\bar{O}_3$
8	$\bar{O}_3$
7	$\bar{O}_3$ and $\bar{O}_2$
6	$\bar{O}_2$
5	$\bar{O}_2$ and $\bar{O}_1$
4	$\bar{O}_1$
3	$\bar{O}_1$ and $\bar{O}_0$
2	$\bar{O}_0$

An external AND gate is required for odd divisions. Reset (active high) puts all outputs high.

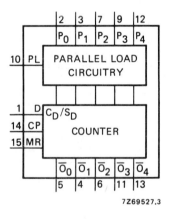

7Z69527.3

4019 Quad two-input multiplexer

Selects one (or OR function of both) four-bit input word. Select A high, B low; word A selected. Select A low, B high; word B selected. Both select lines high, output is A OR B. True data output.

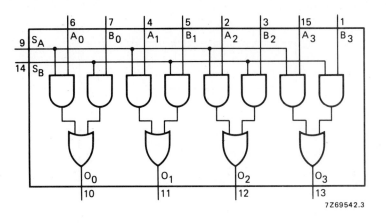

7Z69542.3

4020 14-stage binary ripple counter

Up ripple counter, glitches may be present on decoded outputs. There are no outputs on second and third stages. Counter steps on negative clock edge. Reset active high puts all outputs low and overrides clock.

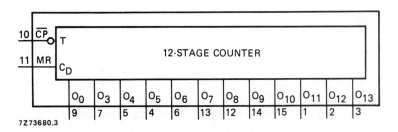

7Z73680.3

398

4021 Eight-bit shift register with parallel load

One-direction shift only from 0-7 (D_S shifts into O). Only outputs 567 are brought out. Register shifts on positive clock edges if load input is low. Data parallel-loaded by taking load input PL high. Load overrides the clock.

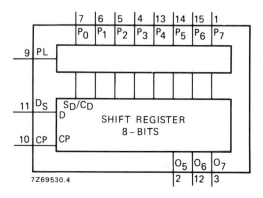

4022 Octal counter/divider

The counter is based on a four-stage Johnson counter with decoded active high outputs. Counter advances on positive clock edges on CPO (with CPI low) or negative edges on CPI (with CPO high). The cascading output is a symmetrical output that is low for counts 4 to 7. Reset is active high, overrides the clock and puts outputs O and cascades out high.

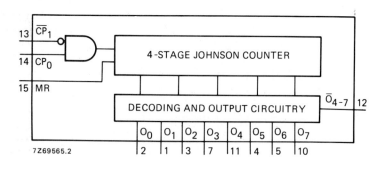

4023 Triple three-input NAND

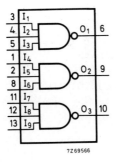

7Z69566

4024 Seven-stage ripple counter
Counter counts up on negative clock edge. Reset active high clears counter and overrides clock.

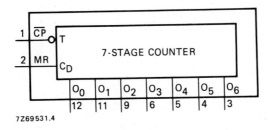

7Z69531.4

4025 Triple three-input NOR

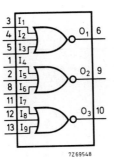

7Z69548

4026 Synchronous decade counter with seven-segment decoded output

Available current 1 mA at 5 volts and 5 mA at 10 volts, so external drivers required for most displays. Outputs true, i.e. HIGH for segment ON. Display has tails on 6 and 9. Normal counting occurs on positive clock edges with reset and clock enable both low. Reset high puts counter to zero and overrides clock. Clock enables high inhibits counting. A low on the display enable input blanks the display. The display enable output is a buffered version of the enable input. Two count outputs are available; a symmetrical divide-by-ten for cascading and a TWO output (denoted C) which goes low on a count of two (used for time of day divide-by-24 and divide-by-60 counter/displays).

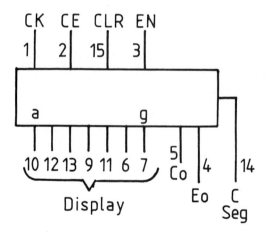

4027 Dual JK flip-flop

Two independent positive edge triggered JK flip-flops. Preset and clear active high override clocked inputs.

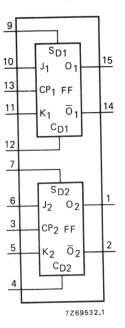

7Z69532.1

4028 BCD decimal decoder

True (active high) outputs. Binary inputs ten through fifteen produce outputs of 8 (even) or 9 (odd).

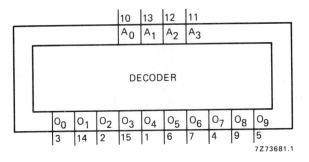

7Z73681.1

4029 Universal up/down binary/BCD counter

For normal counting, load and enable are low. Count mode determined by pin 9; high for four-bit binary counter, low for BCD counting. Counter steps on positive clock edge. Direction determined by pin 10; high for up, low for down. The direction should only be changed when the clock is high.

Carry out $\overline{TC}$ is low for a count of 9 (BCD) or 15 (Binary). This can be linked to the enable of succeeding counter for multi-stage counters. Enable high inhibits counting. Parallel load performed by taking load input high; load overrides clock.

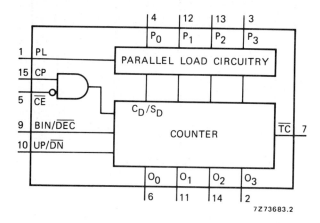

7Z73683.2

4030 Quad exclusive OR gate

Early versions had low input impedance. Other XOR gates: 4070 (different pinning) or 4507.

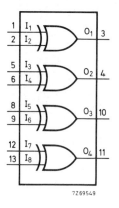

7Z69549

4031 Static 64-bit shift register

Serial-in, serial-out only; used as a store or delay. Shifts on positive clock edge. Input data from input A or B (selected by A/B) allowing a recirculating store to be made by linking O63 to one input, and input data to the other. CO is a buffered output following the clock input.

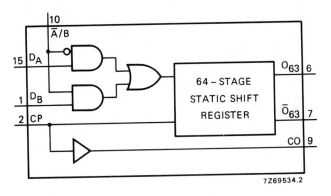

4032 Triple serial adder

Three identical serial adders with internal carry delayed by one clock pulse (see section 8.4). Carry reset is active high and used to clear carry memory before additions. Data A, B presented as two serial words, LSB first. Invert input (active high) complements the output for subtraction and similar applications. Circuit steps on positive clock edge. (The functionally identical 4038 steps on the negative clock edge.)

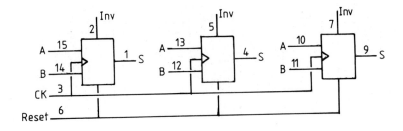

4033 Synchronous decade counter with seven-segment decoder
Functionally similar to 4026 with lamp test and ripple blanking.
Lamp test is active high. If RB1 is low, display is blanked for zero.
RB0 is low if RB1 is low and the count is zero. By linking RB0 to
RB1 of next LS stage leading zero blanking is obtained.

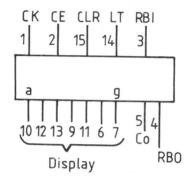

4034 Bidirectional eight-bit bus transfer parallel/serial converter
Eight-bit words A, B can be used as inputs or outputs with controlled
data transfer. Direction is controlled by A/B input; high for A to B,
low for B to A. Transfer can be asynchronous (A/S high) or synchronous on positive clock edge (A/S low). A outputs are tristate and
can be disabled by taking AE input low. AE does not affect loading
of data from the B word. For all the parallel data operations above,
the P/S input must be high. With P/S low, serial data is shifted-in on
positive clock edges to appear at the A or B words as selected by
A/B. The A/S input must be low for serial loading.

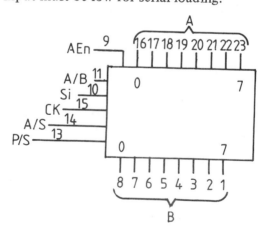

4035 Parallel in/parallel out shift register

Synchronous parallel load. Load input active high, data loaded on positive clock edge. Asynchronous reset active high overrides clocked inputs. Shifts one direction (0 to 3), serial into O. Shifts on positive clock edge. $J\overline{K}$ serial inputs, link for normal serial input. True/complement output controlled by T/C. (High for true.)

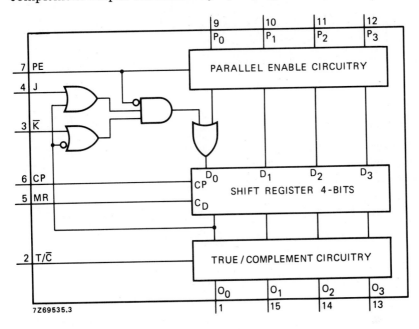

4036 4 × 8-bit RAM

4037 Triple AND/OR

Not commonly available.

4038 Triple serial adder

Function and pinning identical to 4032 except circuit steps on negative clock edge.

4039 4 × 8-bit RAM

4040 12-bit ripple counter
Reset active high overrides clock. Counter steps on negative clock edge.

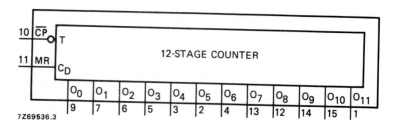

7Z69536.3

4041 Quad true/complement TTL buffer
Complement output will sink 1.6 mA (normal TTL fan-out 1). True output will sink 3.2 mA (TTL fan-out 2). Use 4049 or 4051 for TTL interface for supplies other than 5 volts.

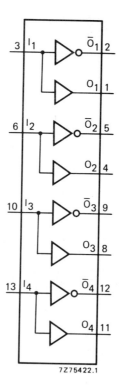

7Z75422.1

4042 Quad hold/follow latch

Four hold/follow latches, *not* D-types. Operation controlled by common polarity and clock inputs as below:

E0 Clock	Polarity E1	Function
L	L	Follow
H	L	Hold
H	H	Follow
L	H	Hold

The circuit follows when clock and polarity are the same.

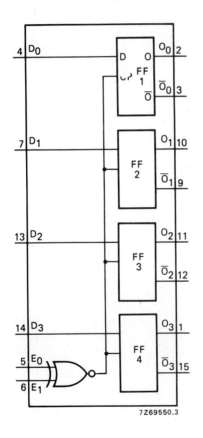

7Z69550.3

4043 Quad RS flip-flops
Set/reset inputs active high (NOR-based flip-flops). Tristate outputs enabled by pin 5; low to disable outputs. Not same pinning as 4044.

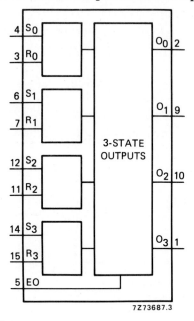

4044 Quad RS flip-flops
Set/reset inputs active low (NAND-based flip-flops). Tristate outputs enabled by pin 5; low to disable outputs. Not same pinning as 4043.

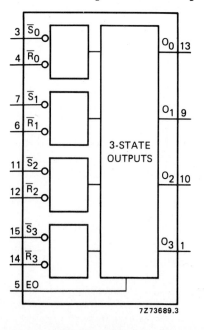

4045 21-stage ripple counter
Not commonly available.

4046 Phase-locked loop

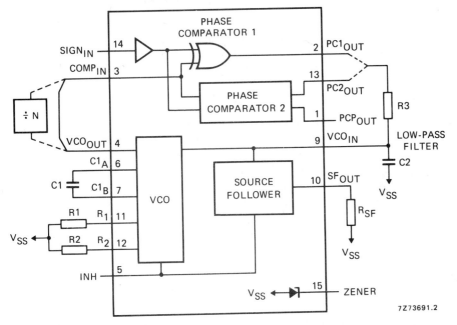

4047 Monostable/astable multivibrator
See section 5.8.2 for full description.

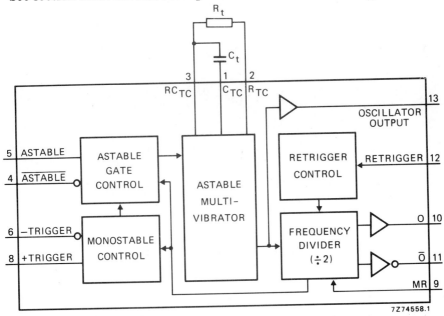

4048 Expandable eight-input gate

4049 Hex inverting buffer
Note that the positive supply is on pin 1 OV on pin 8. This device is designed to convert from normal CMOS levels (10 to 15 V) to TTL. Normally the supply is 5 volts. The inputs will accept voltages up to 15 volts without damage. Outputs can sink 3.2 mA (2 Standard TTL loads, 9 LS loads).

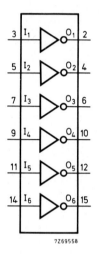

4050 Hex non-inverting buffer
Details as for 4049, but non-inverting outputs.

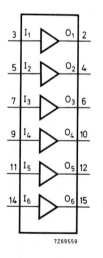

4051 Eight-input analog multiplexer/demultiplexer

Address lines A_0-A_2 select one analog input which is connected to output Z. Note the two negative rails Vss and Vee. Vss is logic ground. Analog signal must be between Vdd and Vee. Vdd-Vee must not exceed 15 volts, but Vee can be negative. ON impedance is 120 ohms. Circuit can work in reverse, i.e. Z can be considered as an input which is connected to one of eight outputs. Enable is active low; if high, all switches are off.

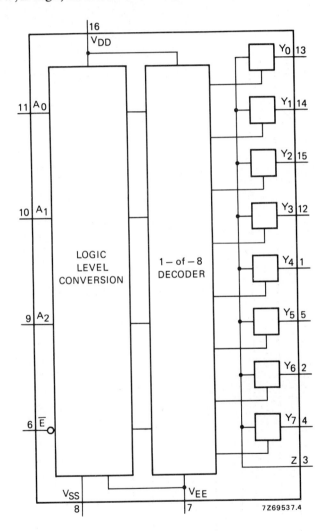

412

4052 Dual 1-of-4 analog multiplexer/demultiplexer
Technically as 4051, except address lines A_0, A_1 select one of four
switches.

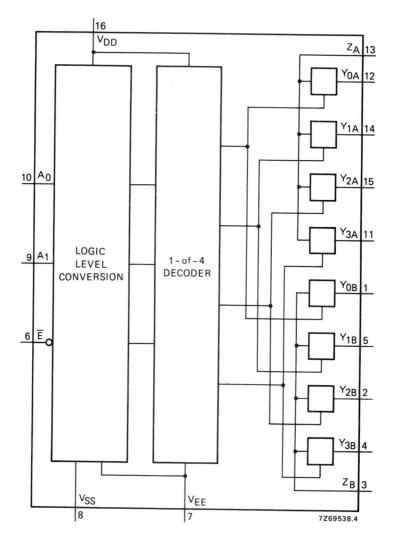

7Z69538.4

4053 Triple two-input analog multiplexer
Functions as three independent changeover switches controlled by
A, B, C. With logic input high, Z is connected to the 1 input; with
logic input low, Z is connected to the 0 input. ON impedance is 120
ohms. Common enable is active low; when high, all switches are
disabled. Vee, Vss as for 4051.

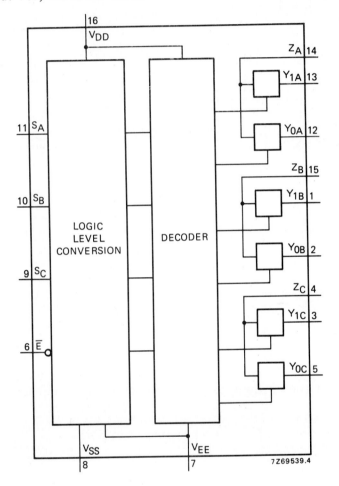

4054 LCD driver
Not available in B version.

4055 BCD seven-segment driver
Not available in B version.

4056 BCD seven-segment driver
Not available in B version.

4057 Four-bit arithmetic unit

4058
Not used.

4059 Programmable counter
Not available in B version.

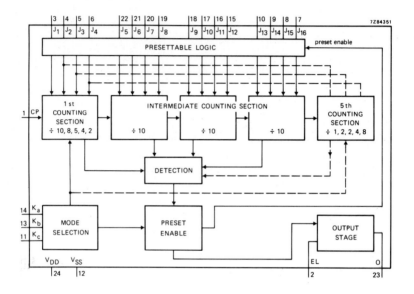

4060 14-bit counter/divider with oscillator
Integral oscillator can function as crystal oscillator, RC oscillator or Schmitt trigger input, as shown. RC oscillator frequency is given by:

$$f = \frac{1}{2.3\ RC}$$

with R in the range 1K to 1M, and C any value above 100 pF. Note that Q2 to Q3 are not brought out. Reset active high clears counter and inhibits oscillator.

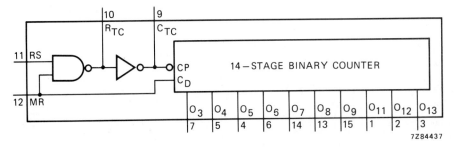

7Z84437

4061 256-bit RAM

4062 200-stage shift register

4063 Four-bit magnitude comparator
Compares two four-bit words A, B to give A<B, A = B, A>B. Cascade inputs provided which are obtained from the outputs of the next less significant device. Cascade inputs on least significant comparator are connected pin 2 low, pin 3 high, pin 4 low. See also 4585.

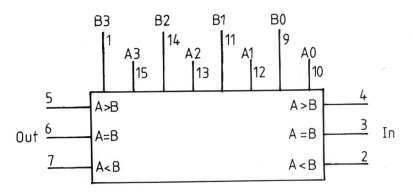

4064
Not used.

4065
Not used.

4066 Quad analog switch (transmission gate)
Identical to 4016 with low (90 ohm) ON resistance.

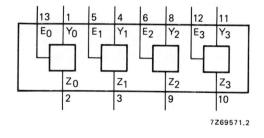

7Z69571.2

4067 One-of-sixteen analogue multiplexer/demultiplexer
Address lines A_0-A_3 select one analog switch. Can function as selecting one of 16 inputs to output Z, or selecting one output for input Z. ON impedance is 200 ohms. Enable is active low; if high, all switches off.

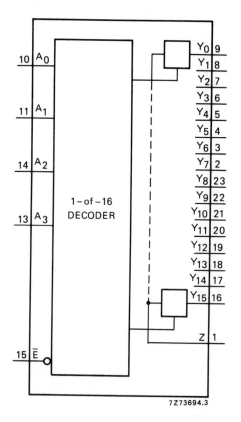

7Z73694.3

4068 Eight-input NAND gate

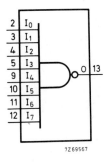

4069 Hex inverter

Note that this device is unbuffered

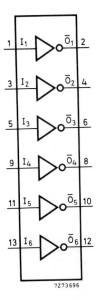

4070 Quad exclusive OR

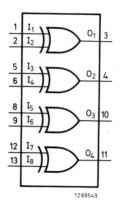

7Z69549

4071 Quad two-input OR

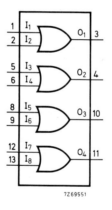

7Z69551

4072 Dual four-input OR

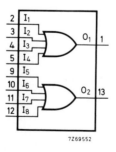

7Z69552

4073 Triple three-input AND

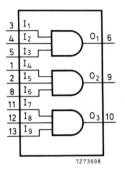

4074
Not used.

4075 Triple three-input OR

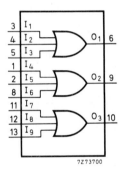

420

4076 Quad D-type with tristate output
True D-type (*not* hold/follow latch). Input enables active low; if either is high, outputs are unaffected by clock. Outputs change on positive clock edge. Output enables are active low; if either is high, outputs are disabled. Data loading can be performed regardless of output enables. Reset active high overrides clocked inputs.

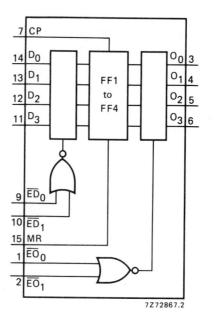

7Z72867.2

4077 Quad exclusive NOR

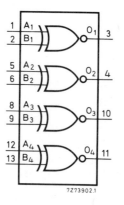

7Z73902.1

4078 Eight-input NOR

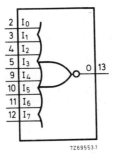

7Z69553.1

4079, 4080
Not used.

4081 Quad two-input AND

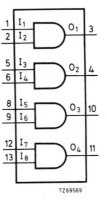

7Z69569

4082 Dual four-input AND

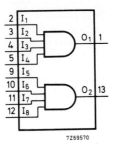

7Z69570

4083, 4084
Not used.

4085 Dual 2 × 2 AND-OR-INVERT gate

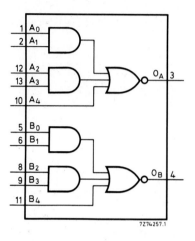

4086 Expandable A-O-I gate

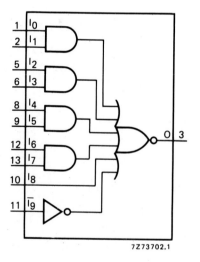

4087, 4088
Not used.

4089 Binary rate multiplier

4090-4092
Not used.

4093 Quad two-input NAND Schmitt
UTP 2.9 volts (5 volt supply) 5.9 volts (10 volt supply).
LTP 0.6 volts (5 volt supply) 2 volts (10 volt supply).

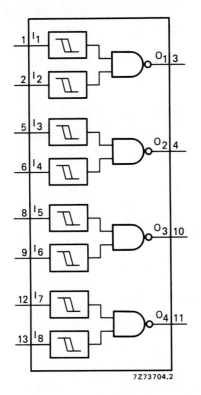

7Z73704.2

4094 Eight-bit shift register and latch with tristate output
Data entered on D, shifted on positive clock edges. Only final stage outputs are brought out directly; output OS appears on positive clock edge and output OS$'$ appears on negative clock edge. Storage registers are hold/follow latches; follow when STR high, hold when STR low. Output enable active high places latch contents on output line. When EO is low, outputs go to high impedance state.

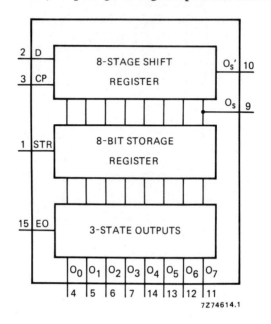

4095 Gated JK flip-flop
Not commonly available.

4096 Gated JK flip-flop
Not commonly available.

4097 1-of-8 analog multiplexer/demultiplexer

Common address lines ABC select one analog switch in each bank. Can select one of eight inputs to output X, Y, or select one output for input X, Y, per bank. ON impedance is 200 ohms. Enable is active low; if high, all switches are off.

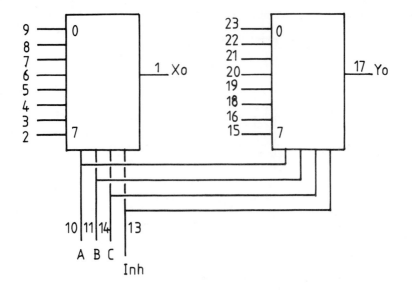

4098 Dual monostable multivibrator

Retriggerable on positive edge (+TR) or negative edge (−TR). Unused +TR should be high, unused −TR should be low. Link Q to +TR or Q̄ to −TR for non-retriggerable operation. Period RC seconds. Reset active low terminates output and inhibits triggering. See 4528, 4538, which are more modern devices with identical pinning.

4099 Eight-bit addressable latch

Data at D input is stored in latch addressed by ABC if Write input is low. Addressed latch follows input (*not* true D-type). Reset active high. If reset high and Write low, all latches except addressed latch are cleared (addressed latch still follows D). If reset and Write are both high, all latches are cleared.

4104 Low to high voltage translator

Translates TTL voltage levels to CMOS levels running on rails other than 5 volts. V_{DD1} is input supply (usually 5 volts). V_{DD0} is output supply. V_{SS} is common 0 V. Output data available in true and complement form. Tristate outputs, enable active high (at *input* voltage levels), low on enable puts outputs to high impedance state. V_{DD1} Pin 16 V_{DD0} Pin 1 V_{SS} Pin 8.

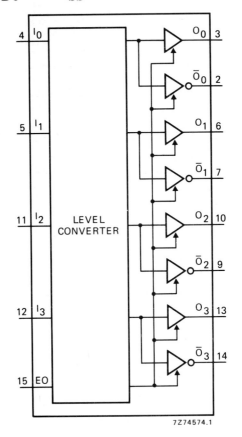

7Z74574.1

Note: Devices in the range 4100-4199, with the above exception are usually given numbers 40100-40199, and as such are listed after the 45XX series devices below. The 4160 BCD counter, for example, will be found as 40160 below.

4502 Strobed tristate hex inverting buffer
Normal enabled operation with $\overline{EO}$ input low and $\overline{E}$ input low; outputs are complements of inputs. With $\overline{EO}$ low and $\overline{E}$ high all outputs are low regardless of inputs. With $\overline{EO}$ high all outputs are disabled (high impedance) regardless of inputs or $\overline{E}$. Outputs can drive two Standard or six LS TTL loads.

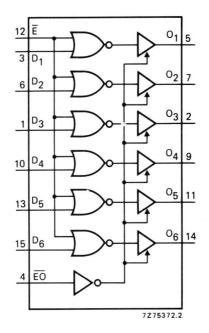

7Z75372.2

4503 Tristate non-inverting hex buffer
Arranged as a group of four and a group of two with separate enables. Enables are active low (high disables outputs). Outputs can drive two Standard or six LS TTL loads.

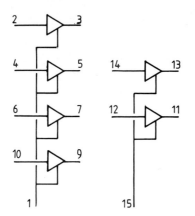

428

4504
Not used.

4505 64-bit RAM

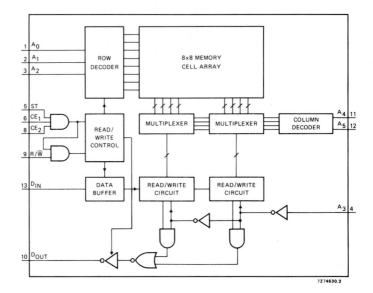

4506, 7
Not used.

4508 Dual four-bit latch with tristate outputs
Hold/follow latches *not* true D-types. With Store input high, outputs follow inputs. With Store input low, outputs hold. Clear active high clears latches and overrides inputs. Output enable active low (high disables outputs). Output enable does not affect hold/follow operation.

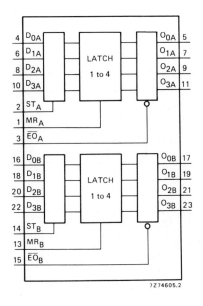

7Z74605.2

4509
Not used.

4510 BCD up/down counter
For normal counting carry in CE, reset MR, load PL are low. U/$\overline{\text{D}}$ input is high for count up, low for count down. Counter steps on positive clock edge. Reset active high overrides clock. Asynchronous parallel load achieved by taking load input high. Stages cascaded by linking carry out to carry in of next more significant stage. Carry in high inhibits counting.

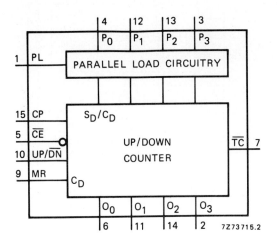

7Z73715.2

4511 Four-bit hold/follow latch with seven-segment decoder/driver
Store input $\overline{EL}$ low, latches follow input. Store input high, outputs
hold. Latch outputs are not brought out. Outputs (high for segment
on) can source 25 mA but sink only normal CMOS currents. Blanking, active low, extinguishes all segments. Invalid BCD input extinguishes all segments. Lamp test, active low, lights all segments. Lamp
test overrides blanking.

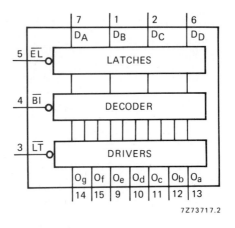

7Z73717.2

4512 Eight-channel data selector (multiplexer) with tristate output
Address S_0-S_2 selects one input, which is routed in true form to the
output. Enable input active low gates output; output is low if enable
input is high. Tristate output is controlled by Eo. Output enabled for
Eo low, disabled for Eo high. This device is not an analog multiplexer.

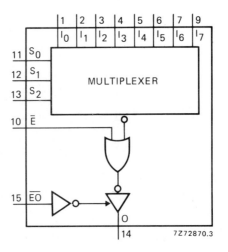

7Z72870.3

4513 BCD to seven-segment decoder/driver with hold/follow latch
Functionally as 4511 with ripple blanking. Not commonly used.

4514 Four-bit latch/decoder
Hold/follow latch, not true D-type. EL high, follow; EL low, hold.
True outputs, i.e. selected output is high, others low. Output enable
active low (high input puts all outputs low). State of output enable
does not affect hold/follow operation.

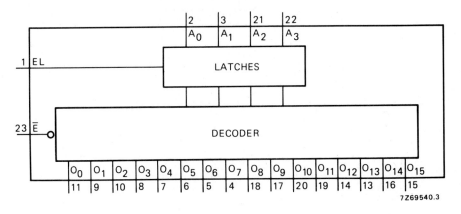

4515 Four-bit latch/decoder
As 4515 functionally and pinning, but complement output. High
input on output enable puts all outputs high.

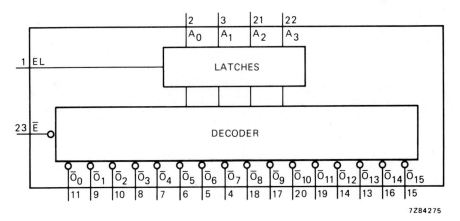

4516 Synchronous binary up/down counter
U/D controls direction; low for down, high for up. Counter steps for clock positive edge. Carry in $\overline{CE}$ acts as clock inhibit; low for counting high for inhibit. Counters can be cascaded by linking $\overline{TC}$ to $\overline{CE}$ of next more significant stage. Load active high (low for normal counting). Reset active high. Clock must be low for load or reset.

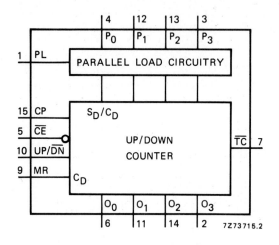

7Z73715.2

4517 Dual 64-bit shift register

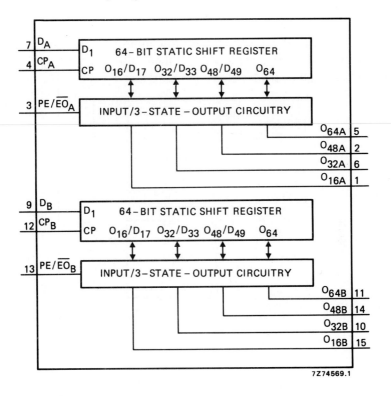

7Z74569.1

434

4518 Dual synchronous BCD counter

Two identical up counters which can be made to step on positive or negative clock edges as follows:
Positive edge on CPO, CPI high (CPI low inhibits counting).
Negative edge on CPI, CPO low (CPO high inhibits counting).
Reset active high overrides clock.

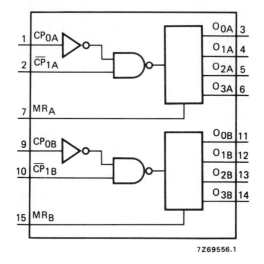

7Z69556.1

4519 Quad two-input multiplexer

Pinning and function as 4019 except that if both select lines are high the XOR of the A, B words is produced.

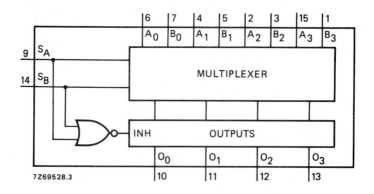

7Z69528.3

4520 Dual synchronous binary counter
Pinning and function as 4518 except four-bit binary count.

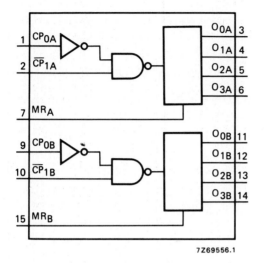

7Z69556.1

4521 24-stage frequency divider

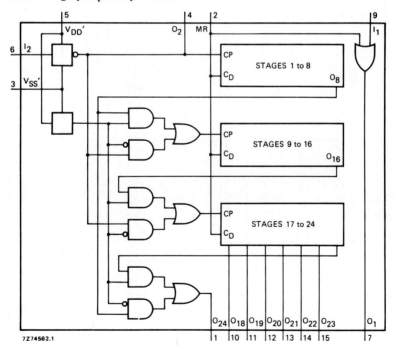

7Z74562.1

4522 Programmable BCD counter

A four-bit BCD *down* counter with parallel load and facilities for detecting zero count. Counter steps on positive edge of CPO or negative edge of CPI (see function of 4518 above). Parallel load input PL active high overrides clock. For single-stage programmable divide-by-N, N is set on parallel inputs and active high TC (high on zero count) linked to PL. CF is cascade input and must be high for TC to go high. Schmitt trigger clock inputs. Reset active high, overrides clock and PL.

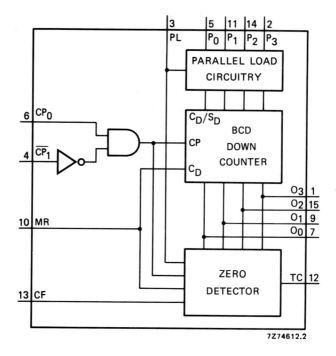

7Z74612.2

4523-4525
Not used.

4526 Programmable binary counter
Function and pinning as for 4522 but four-bit binary down counter.

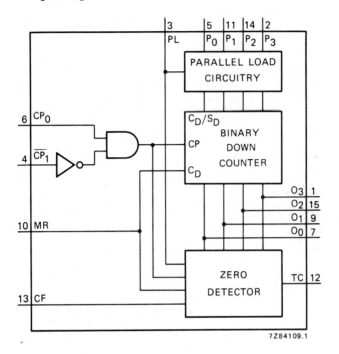

7Z84109.1

4527 BCD rate multiplier

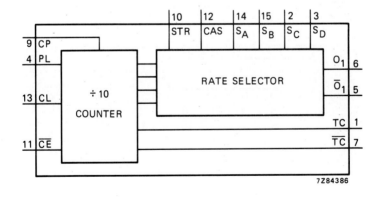

7Z84386

4528 Dual resettable monostable
Two identical monostables with retrigger and reset facilities. Identical in function and pinning to 4098.

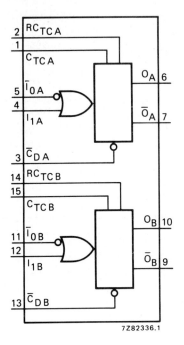

7Z82336.1

4529 Dual four-input analog multiplexer

4530 Dual five-input majority vote
Output takes state of the majority of inputs.

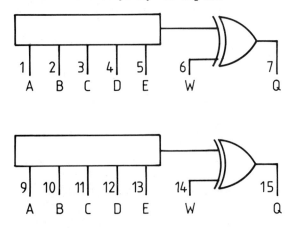

4531 13-bit parity tree

Output is low if an even number of inputs are high. For less than 13 bits, device can be used to generate even or odd parity by using one input as sense select.

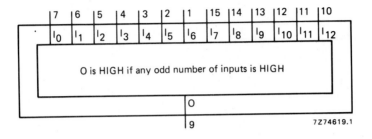

4532 Eight-level priority encoder

The device has eight inputs active high. The three outputs O_0, O_1, O_2 indicates the binary number equivalent to the largest high input (e.g. if 0, 1, 2, 6 inputs were high, output would be binary 6, 110). GS is high if *any* input is present. Cascade output Eo is high if *no* inputs are high and Ein is high. Enable input is active high; when low, *all* outputs are low.

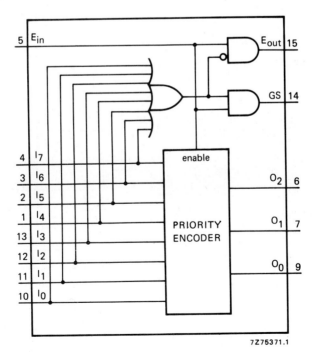

4533

Not used.

4534 Five-decade counter

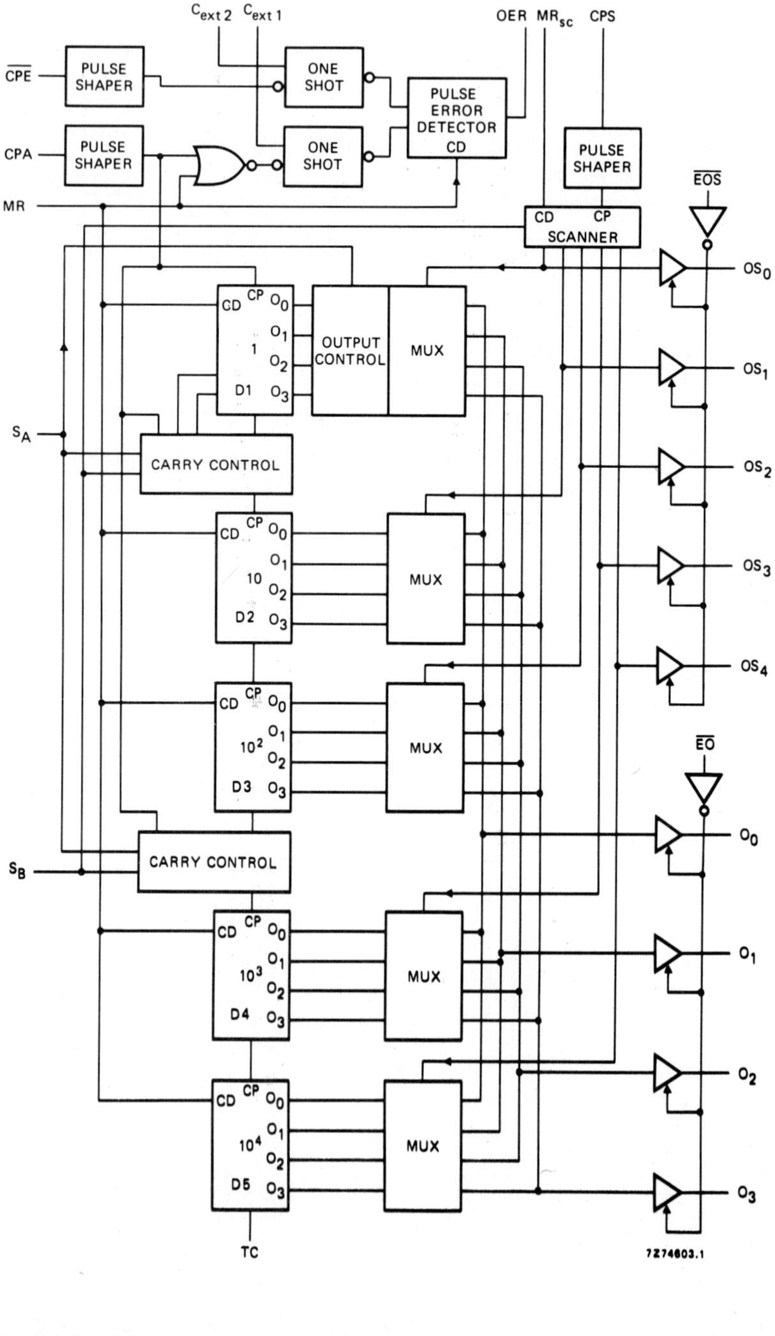

4535
Not used.

4536 Programmable timer

4537
Not used.

4538 Dual retriggerable monostable multivibrator
Pinning and function as for 4528 with Schmitt trigger inputs.

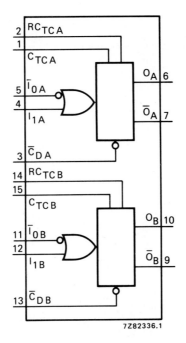

7Z82336.1

4539 Dual four-channel data selector/multiplexer

Common address S_0, S_1 selects one of four inputs in each section. True output data. Enable active low (high gives permanent low output).

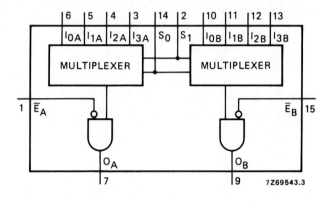

7Z69543.3

4540
Not used.

4541 Programmable timer
See section 5.5.3 for full description.

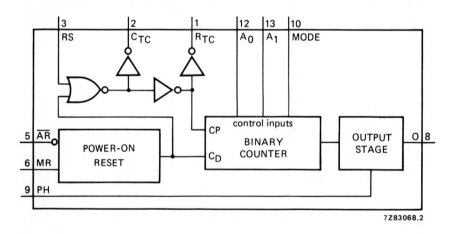

7Z83068.2

4542
Not used.

4543 BCD to seven-segment latch/decoder driver

Hold/follow input latches controlled by LD; High, follow; Low, hold. BL is active high blanking input (high blanks display). Phase PH controls sense of output; low gives high output for segment on, high gives low output for segment on. For LCD displays, connect as shown and drive PH from oscillator. Can source or sink 10 mA. Non-BCD input blanks display. 6 and 9 have tails.

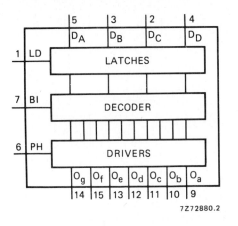

7Z72880.2

4544-4550
Not used.

4551 Quad two-input analog multiplexer/demultiplexer
Functions as four changeover switches controlled by A. With A high, Z is connected to 1 input; with B low, Z is connected to 0 output. ON impedance is 120 ohms. V_{EE}, V_{SS} as for 4051.

4552
Not used.

4553 Three-digit BCD counter with multiplexed output

4554 Binary multiplier (2 × 2)

4555 Dual 1-of-4 decoder/demultiplexer

Independent address lines A_0, A_1 select one of four active high outputs. Enable active low (when high all outputs are low). Enable can be used as demultiplexer data input, selected output is the complement of enable input.

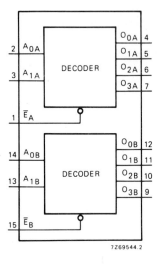

4556 Dual 1-of-4 decoder/demultiplexer

Function and pinning as for 4555 but outputs active low.

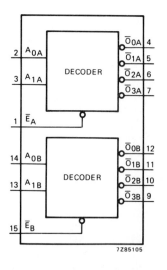

4557 64-bit shift register

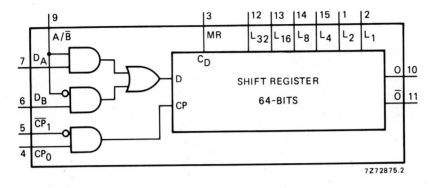

4558, 4559
Not used.

4560 BCD adder
Performs true BCD addition on two four-bit BCD numbers A, B.

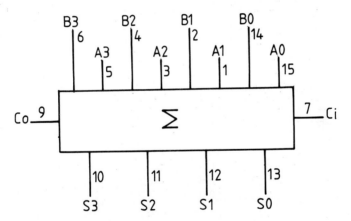

4561-4565
Not used.

4566 Timebase generator

4567-4579
Not used.

4580 4 × 4 multiport register

4581 Four-bit arithmetic unit

4582 Look ahead carry for 4581

4583 Dual Schmitt with adjustable threshold

4584 Schmitt trigger
UTP and LTP as for 4093.

4585 Four-bit magnitude comparator
Compares two four-bit words to give three results: $A > B$, $A = B$, $A < B$. Outputs are active high. Cascade inputs are available for comparison of more than four-bit words. Connect = and < outputs to corresponding inputs on next more significant stage. Connect all > inputs high; < inputs on least significant stage low > = on least significant stage high.

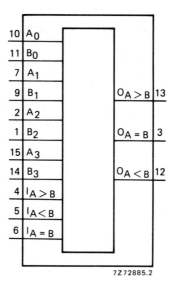

7Z72885.2

40097 Tri-state hex buffer
Not to be confused with 4097. True data output. Independent 4 and 2 groups. Enables active low (high gives high impedance output state). Will drive one Standard TTL load.

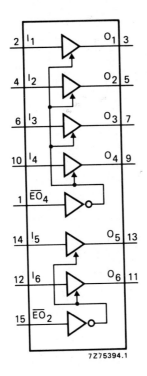

7Z75394.1

40098 Tristate hex inverting buffer
Not to be confused with 4098. As 40097, but inverted data output.

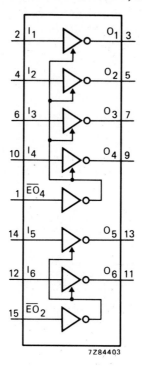

7Z84403

40106 Hex Schmitt trigger

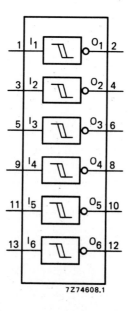

7Z74608.1

40160 Synchronous four-bit decade counter
Count, parallel load all synchronous on positive clock edge. PE active
low overrides count. To count, both CET and CEP must be high.
Count up only. TC is cascade output and is linked to CET on next
more significant stage. Asynchronous reset MR active low overrides
clock, CP.

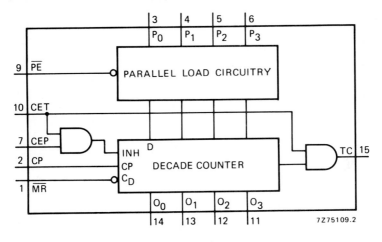

40161 Synchronous four-bit binary counter
As 40160, but binary counter. Pinning identical.

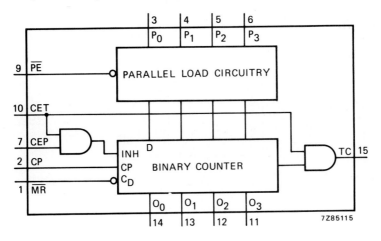

40162 Fully synchronous four-bit decade counter

As 40160, but reset is synchronous active low. Reset overrides parallel load overrides count. Pinning identical.

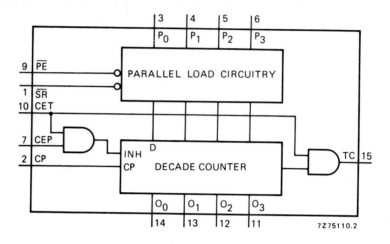

40163 Fully synchronous four-bit binary counter

As 40162, but binary counter. Pinning identical.

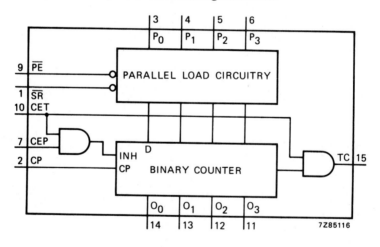

Index

About the Author

E. A. Parr is an Electrical Engineer at Sheerness Steel. He has written many articles in the electronics press, and is the author of several other British electronics books including *OP AMP Handbook, Electronics Pocket Book,* and *Beginner's Guide to Microprocessors*.